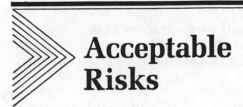

Acceptable
Risks

Also by Pascal James Imperato, M.D.
A Wind in Africa
What to Do About the Flu
African Folk Medicine
Medical Detective
The Administration of a Public Health Agency

Also by Greg Mitchell
Truth . . . and Consequences

Acceptable Risks

Pascal James Imperato, M.D.
Greg Mitchell

Viking

VIKING
Viking Penguin Inc., 40 West 23rd Street,
New York, New York 10010, U.S.A.
Penguin Books Ltd, Harmondsworth,
Middlesex, England
Penguin Books Australia Ltd, Ringwood,
Victoria, Australia
Penguin Books Canada Limited, 2801 John Street,
Markham, Ontario, Canada L3R 1B4
Penguin Books (N.Z.) Ltd, 182–190 Wairau Road,
Auckland 10, New Zealand

First published in 1985 by Viking Penguin Inc.
Published simultaneously in Canada

LIBRARY OF CONGRESS CATALOGING IN PUBLICATION DATA
Imperato, Pascal James.
  Acceptable risks.
  Bibliography: p.
  Includes index.
  1. Medicine, Preventive.   2. Safety education.
3. Health.   I. Mitchell, Greg.   II. Title.
RA431.I46   1985   613   83-40634
ISBN 0-670-10205-9

Printed in the United States of America
by The Book Press, Brattleboro, Vermont
Set in Melior
Designed by Ann Gold

For
Sydney Downey
who would have enjoyed it
                    PJI

and

For
Stanley Mitchell
          GM

Acknowledgments

We would like to thank William Strachan of Viking Penguin for first suggesting this book and for his editorial assistance. We also want to thank Virginia Barber for her interest and help. We are grateful to the numerous individuals who gave us the benefit of their expertise and knowledge; special thanks go to Allen D. Spiegel, Ph.D., of the State University of New York, Downstate Medical Center, and the staffs of the Library of the New York Academy of Medicine, National Library of Medicine, New York Public Library, and Downstate Medical Center Library for assistance in locating primary source material. And to our wives, Eleanor and Barbara, we express our gratitude for patience and assistance. We are grateful to Phyllis Alexander, Kay Dobson, Molly Kessert, Constance Jones, Cheryl Moore, and Maureen Roaldsen for their careful preparation of several drafts of the typescript.

Contents

Introduction

We live in a dangerous world. Asbestos is in our schools, chemicals in our food, and drunk drivers on our highways. What were once staples are now vices: salt makes us hypertensive; whole milk causes clogged arteries. If you live your entire life in a major American city your chances of being murdered are greater than the threat of death in combat for American soldiers in World War II. In the age of herpes and AIDS, even sex is risky.

During this century scientists and physicians have conquered nearly all of the classical scourges afflicting Americans—diphtheria, tuberculosis, malaria, typhoid, smallpox, polio, rickets, and pellagra—and we are living longer, from an average of fifty years in 1900 to seventy-five today. Pollution controls are up and infant mortality down. New hazards, however, have taken the place of those eradicated: pesticides in our food, chemicals in the air and water (and practically everything else), guns in the hands of thousands of teenagers. Each of us faces a three in ten chance of eventually dying of cancer. And no matter how successfully we manage to eliminate or avoid the dozens of risks of day-to-day life, we might at any moment, awake or asleep, perish in an exchange of even a fraction of the world's fifty thousand nuclear warheads.

Yet we go on risking our lives, and for good reason: there is no alternative. Risks are ubiquitous.

Some of us tend to be risk-takers, others try to avoid risks religiously, but nearly all of us accept the variety and prevalence of risks in our lives and only rarely stop to think about any given risk for very long. A recent Department of Transportation study, for example, found that nearly half of the motor bridges in the United States are structurally deficient or obsolete—yet how many people worry about bridge collapse every time they cross a highway span? According to nearly every expert, it is now likely that Southern California will experience a cata-strophic earthquake some time in the next thirty years, yet the population of the areas most at risk continues to soar.

On what could be (but was not) called "Black Friday," March 2, 1984, the National Research Council informed the American public, following a three-year study, that tens of thousands of chemicals widely used in commercial products have never been adequately tested for potential health hazards. This includes 64 percent of the ingredients in drugs, 66 percent of those in pesticides, 84 percent in cosmetics, and 81 percent of food additives. John C. Bailair, a health statistician who headed the committee's panel on sampling, said he suspected the results would "surprise a lot of people" who thought chemicals had been closely scrutinized by regulatory agencies.

But were we really surprised? From the lack of public or political outcry in the wake of these revelations it certainly didn't seem so.

Accepting risks has become second nature to us. As Margaret Mead pointed out, life is made possible by the ability to "forget pain." Not many insist on a zero-risk society. But acceptability is subjective, danger a matter of degree.

We choose or embrace so many risks ourselves—from smok-ing to speeding on the interstate—it's hard to point our fingers at others who seem to have it in for us. With so many threats to our health and well-being imposed on us it is surprising that we freely choose to adopt so many others.

The problem with so-called "voluntary" risks, however, is that one person's chosen risk has a way of becoming another person's imposed risk—as thousands of victims of drunk driv-ers find out every year. Indeed, alcohol abuse, a chosen risk, has been linked to half of all automobile accidents, half of all homicides, one-quarter of all suicides, and 40 percent of problems brought to family court. Its annual economic cost to the American society has been estimated as high as $120 billion.

In addition, some chosen risks are chosen more freely than others because most people are uninformed—or misinformed—about many of the risks they face. Information on risks is often inadequate, overly complicated, biased to suit vested interest, or just plain wrong.

In other cases, however, we receive good advice, but reject it, or choose to ignore it entirely; instant gratification overwhelms good sense. "The fundamental question is whether people who are given more knowledge will change lifelong behavior patterns and act in a safer way," says R. David Pittle, a former member of the U.S. Consumer Products Safety Commission, and now a director of the Consumers Union. "The majority of public campaigns designed to change people's behavior permanently simply have not worked."

Sometimes we get to "choose our poison," but other times it chooses us: risks are imposed by others. In many instances we know what's going on and can take evasive action, but in other instances, industry officials—or our government "watchdogs"—could or should have prevented the exposure in the first place. The fact that we embrace certain risks does not give others carte blanche to impose quite different risks on us. We have shown that we can be quite reasonable about accepting our fair share of abuse in return for living in a modern industrial society if we have been given an adequate explanation as to why this is necessary. But we hate negligence and bristle at risk-for-profit; we don't like being taken advantage of.

Obviously there are degrees of risk and also degrees of what is called "safety." Whether the defined risk is viewed as acceptable or not is a complex matter influenced by many factors, ranging from the magnitude of the risk to personal value judgments. Usually the decision comes down to: "Is *this* risk worth taking?" Inevitably, we construct personal cost-benefit equations—sometimes unconsciously, in a matter of seconds—before embarking on risky behavior.

But what may prove useful when employed personally in an informal, instinctive way is increasingly being utilized systematically by government bureaucrats and industry analysts to justify the introduction of new risks (and the continued existence of old ones). The terminology is benign—who could quarrel with "risk assessment," "risk analysis," or "risk management"?—but the results are often controversial. Elaborate formulas compare the benefits of reducing certain risks with the costs of accomplishing that risk reduction.

With risks proliferating and a finite pool of resources available to counteract them, there certainly is a place for cost-benefit analysis in our society, but many critics argue that cost-benefit analysis is fraudulent at its heart because many benefits cannot be measured at all. (The "benefit" of escaping the agony of emphysema may seem priceless, especially to those suffering from it.) As Richard Ayres of the Natural Resources Defense Council said, referring to cost-benefit aficionados: "They are trying to put into numbers something that doesn't fit into numbers, like the value of clean air to our grandchildren."

This hasn't prevented some analysts from putting a price on human life itself; recent estimates have attempted to peg that value at somewhere between $49,000 and $1 million, with $200,000 to $300,000 a widely accepted level. This would seem to sell life pretty cheap, especially since many utility infielders in major-league baseball make more than that in a single season. As in so many aspects of the acceptable risk debate, however, this question comes down to personal perspective and what could be called *reasonableness*; what may seem unacceptably paltry to the individual might be all society as a whole can afford. If we're not all $6-million-men and -women, what *are* we worth in regulatory spending? Are we better off sick and affluent than healthy and poor? How much are *you* willing to pay for regulations and measures to keep your *neighbor* alive?

When Ronald Reagan became President one of his first acts was to mandate that cost-benefit analysis was to precede all federal regulatory decision making. The residents of Tacoma, Washington, became involved in the best-known ramification of this new policy.

On July 12, 1983, William D. Ruckelshaus, administrator of the Environmental Protection Agency, asked the citizens of Tacoma to debate the following question: What cancer risk is "acceptable" to keep 575 people working in a local copper smelter that is releasing arsenic into the air? This was to be the first test case in a new federal plan which would involve the public in assessing the acceptability of pollution and other health hazards.

"For me to sit here in Washington and tell the people of Tacoma what is an acceptable risk," Ruckelshaus said, "would be at best arrogant and at worst inexcusable." Ruckelshaus said he wasn't holding a referendum, just looking for a consensus, but he admitted: "I don't know what I'll do if there is a fifty-fifty split."

The Tacoma debate was prompted by a proposed EPA standard on arsenic emissions into the air. Although the new standard would apply throughout the country, it would primarily affect Tacoma because the copper smelter there, operated by Asarco Inc., was the only one that uses ore with a high arsenic content. The EPA said the new standards were more stringent but would not eliminate all health risks. The only other option was to close the plant, with a loss to the local economy of $20 million a year.

This would be a grave consequence—but the health risks were also said to be high. EPA estimated that under the current standards an additional four lung cancer deaths each year among the three hundred sixty thousand people who lived within a twelve-mile radius of the plant could be attributed directly to the arsenic emissions. Under the new standards the result would be at least one extra lung cancer death each year. Those living closest to the plant were most at risk. Even under the new standards they would run a one-in-fifty chance of contracting lung cancer during their lifetime. (Traditionally, federal agencies have considered any risk of cancer greater than one in 100,000 to be unacceptable.)

Looked at from the perspective of society as a whole, that one extra annual death may seem low; from the perspective of that year's victim (and his or her family and friends), of course, it would look quite different. Even so, an informal equation could be set up, in classic cost-benefit terms: was saving one life per year, and an untold amount of illness—by closing the plant—worth 575 jobs and $20 million to Tacomans?

There seemed to be, to EPA, no middle ground. Asarco was

already willing to spend $4.4 million to meet the new standards; but more stringent rules would supposedly make it economically unviable for the plant to stay open.

Response to Ruckelshaus' plan from the national media was generally negative. In an editorial titled "Mr. Ruckelshaus as Caesar," *The New York Times* said it was "inexcusable" for Ruckelshaus to impose an "impossible choice" on Tacoma. The newspaper suggested that it was unconscionable for the agency to ask the public to accept a high cancer rate while indicating that there was no middle ground between that and mass unemployment. Others complained that this "jobs blackmail" threat would lead to a sorry spectacle pitting workers against residents.

But Ruckelshaus held firm. In a speech before the National Academy of Sciences, he had indicted regulation by "emotionalism" and said that risk management (balancing the health benefits of reducing risk with the economic costs of doing so) should be incorporated into a government-wide process of regulation. "We must assume that life now takes place in a minefield of risks from hundreds, perhaps thousands of substances," Ruckelshaus said. "No more can we tell the public: You are home free with an adequate margin of safety."

The battle of Tacoma formed along predictable lines. "If the proposed controls reduce deaths from four to one, that is a very good thing," said Charles O'Donahue, who heads the United Steelworkers of America local which represents the smelter's employees (whose respiratory cancer rate was two and a half times the normal rate). "One death is something we aren't going to cure. Simply dying from cancer is not different from a man losing his job and then committing suicide." The union started a door-to-door educational campaign. Ruth Weiner, a local professor, charged on the other hand that the EPA was "copping out . . . it is up to the EPA to protect public health, not to ask the public what it is willing to sacrifice not to die from cancer."

Tacoma Mayor Douglas Sutherland struck a balance. Putting all the evidence on the table, he said, "gives the residents a chance to make a personal choice—to leave or stay. . . . Decisions that affect our basic lives are always agonizing." He pointed out that workers and environmentalists "want the same thing—

to live in a healthy environment, and still *work*." But the first newspaper poll on the subject found a majority of residents favoring a plant shutdown.

In early November 1983 the EPA held three days of hearings in Tacoma. While a diversity of views from the community and experts were expressed, the overwhelming opinion was that there had to be a way to keep the plant open *and* reduce the risks further than EPA and Asarco seemed willing to accept. Ruckelshaus' plea for a community consensus around one of the two options instead led to a new idea—"community air standards," imposed by the state. While the proposed EPA rules were concerned with emissions at the source, the community standards would set limits on the level of arsenic that could be detected in the air along the perimeter of Asarco's property.

The local residents, essentially, had decided neither to trust, nor to wait for, EPA's decision (which was, indeed, subsequently delayed and had still not been made months after the hearings ended). To ease the burden on Asarco, the state promised to phase in the new standards, but the company still contended that it could not meet the final limits.

And indeed, in June 1984, Asarco announced that it would close its smelter because of low copper prices and the high cost of environmental protection. Five hundred of the smelter's 550 employees would lose their jobs. But on the plus side, the company said that it had (due to environmental pressure) designed a new process to produce arsenic from metallic byproducts which generates no arsenic emissions.

The Tacoma debate indicated two things: First, the federal government is serious about "risk management" in which a difficult (perhaps unworkable) meshing of cost-benefit formulas and common sense replaces idealistic standards. Second, while many posture as purists, most people, in practice, are quite willing to be *reasonable*—to accept some degree of risk for some amount of reward. Whether this involves smoking, driving fast, taking birth control pills, or living in "earthquake country," the psychology is the same: nothing ventured, nothing gained.

The national debate over the nature and extent of government regulations has only just begun. For the remainder of this dec-

ade and during the next, the arguments over where and how
meager risk-cutting resources should be spent will intensify.
High federal deficits preclude lavish risk-cutting expenditures.
Having to choose between risks is nothing new; what charges
the debate today is that the public has become increasingly
aware of risks. We have grown more knowledgeable about the
tragic nature and effects of many risks. As we learn more about
how risks are being "managed" for us, we may not like what
we learn. The debate over costs, benefits, risks, and prevention,
as well as the role of individuals, industry, and government in
setting "acceptable" standards, will escalate. Attempts to define
risk acceptability or unacceptability will be molded in part by
the interplay of social, cultural, political, and economic forces.

In this book we discuss how risks are defined and responded
to on a personal, corporate, and governmental—even global—
basis; why risks become acceptable for some and unacceptable
for others. Clearly there is plenty of blame to go around. We
believe that industry too often knowingly endangers the public.
We think that government should more actively—or at least
more effectively—protect the public. But we never lose sight
of the public's often ambivalent attitude—our love/hate relationship with risks.

Most of the book's eight chapters begin with background and
analysis of a general risk topic (choosing risks, defining risks,
imposing risks, and so forth), followed by illustrative "close-
up" looks at specific risks. These case studies present in-depth
examinations of important, timely, and revealing issues based
on extensive research and personal interviews. Any number of
issues could have been chosen as illustrations. We have selected
issues—such as cholesterol, low-level radiation, and clean air—
that are current, that are vital to a variety of individuals, and
that clearly illuminate the topic under discussion.

We begin by concentrating on what we term *chosen* risks,
situations in which individuals purposely expose themselves
to harm. In this chapter we raise questions about why only one
in eight Americans wears seat belts, why some people smoke,
and why others choose to live in earthquake-prone Southern
California.

From choosing risks we move on to analyze how the public

is informed, misinformed, and uninformed about risks. We look at how even the well informed tend to believe what suits their needs or appeals to their prejudices, and we examine how the acceptance of risk may be the result of misguided judgment applied to misinformation in order to fulfill personal needs.

We next consider imposed risks such as carcinogens and mutagens in foods and unsafe drugs. How does industry justify its excesses? And why has the government often failed to perform its watchdog role? These questions are separately discussed in two subsequent chapters, "Government Guardians" and "Calculating Risks." In "Defining Risks" the difficulties of risk determination as well as determining "acceptability" are discussed in detail.

The book then examines perhaps the ultimate risk—"The Greenhouse Effect"—and in a final chapter, "What You Can Do About Health Risks," we describe ways in which individuals can monitor themselves and their surroundings and reduce risks in everyday life.

Life has always been full of risks, and some in fact have argued that a riskier life is a richer life; dancing on the edge of the abyss can be stimulating. For most people, however, most risks are genuinely unappealing most of the time. We may or may not be in greater peril than our ancestors—we don't die from smallpox but they didn't have to worry about The Bomb—but we are certainly more painfully aware of our vulnerability.

But are we aware enough? If we are, how much should we worry about risks? And are we, and those charged with protecting us from harm, doing enough to minimize risks? Because of misinformation, poor judgment and ignorance, technological progress and corporate greed, the increasing need to balance costs against benefits, the frequent failure of government to protect the public, and our just as frequent failure to protect ourselves, we now live in a world of risks that are respectable—*acceptable risks*.

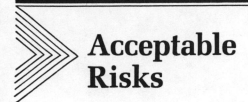

Acceptable
Risks

1
Chosen
Risks

Of the ten leading causes of death in America today, seven could
be substantially reduced if people chose to eat properly, exercise more, not smoke or abuse alcohol, drive safely, and reduce
stress. Consider the first minutes of the day. That first cup of
coffee may contain three ingredients that receive a good deal
of public and medical attention these days: caffeine, whole
milk, and sugar. For breakfast many eat a cholesterol-rich egg
(or two). By the time they leave for work millions of Americans
have already had their first cigarette. When they climb into their
cars for their daily commute, only one in eight will fasten their
seat belts. And the day is only an hour old.

People who knowingly take risks do to themselves what they
wouldn't let others do to them. But we all do this.

Perhaps because we believe what we have been told so often,
and have observed in the course of our own lives: there are no
rewards without risks. Some have gone so far as to declare that
risk-taking for benefit is the essence of human striving.

Consciously or unconsciously we make crucial personal risk
judgments all the time: Does a particular risk exist? To what
degree? What compensation does it carry? And, therefore, is it
acceptable or unacceptable?

There is, obviously, no consensus on acceptability. What is
acceptable is open to judgment, and judgments vary. Although
these judgments are arrived at through the exercise of free will,

they are frequently influenced by a number of outside forces that often function in an extremely subtle manner.

A chosen risk could be defined as one taken with some knowledge of possible harmful consequences. "Taking a risk" does not necessarily imply action, however. Smoking is risky; so is *not* buckling your seat belt. And, in a sense, we choose a risk every time we acquiesce in a decision made by others that permits potential danger.

For each risk the consequences vary, and so does each person's perception of the risk. Yet a high degree of danger, accurately perceived, does not always deter risky behavior. We assemble our personal-risk portfolios in an erratic (and sometimes irrational) way. Our reactions to risks are almost never *entirely* appropriate. We blithely accept a substantial risk (such as taking a long automobile trip in bad weather) but are often inordinately afraid of long-shot threats (such as being hit by lightning).

Most of us are bundles of contradictions. Many people carefully strap on their seat belts and then smoke half a pack of cigarettes on the drive to work. We denounce air pollution but live in smog-bound cities. Some sacrifice culinary enjoyment to cut down their intake of saturated fats but refuse to install $20 smoke detectors in their homes. In essence, we ignore a great many risks to concentrate on avoiding a select few, usually from a short-run, crisis-oriented perspective.

While we can get angry about risks imposed on us, we are philosophical about those we choose ourselves, especially when we feel we cannot easily avoid them. Many Americans choose to live in big cities despite the likelihood that they will eventually become victims of crime. Ultraviolet rays from the sun harm everyone, from wrinkling the skin to causing cancer, but few people avoid the sun (and many worship it). The attitude in these cases is: "What can you do?"

This may keep us sane, but it can also be exploited—as in a government propaganda film of the 1950s, excerpted in the movie *The Atomic Cafe*. It showed a man slipping on a bar of soap in the shower, a "life-is-full-of-risks" message meant to convince us that we should happily accept the new risk of fallout from atomic tests.

One of the most basic psychological reactions is known as denial. While denial may be an appropriate response to sources of anxiety that cannot be eliminated, it is maladaptive when used to deal with threats that can be dealt with. Half of all Americans, for example, think that "everything causes cancer" and "there's not much a person can do to prevent cancer," according to a recent National Cancer Institute survey. This fatalistic attitude is largely uncalled for, and self-destructive, since up to one hundred thousand cancer deaths a year, the Institute noted, could be avoided if people changed some of their living habits that cause cancer.

Life is full of risks. Even peanut butter kills. (According to one computation, people who eat four tablespoons of peanut butter a day have an eight-in-one-million chance each year of dying from cancer caused by aflatoxin, a substance commonly found on nuts.) And everyone chooses—or chooses between—risks.

But some risks are chosen more freely than others: Smoking cigarettes, for instance, may be based on free choice, while undergoing an X ray after suffering a severe injury is almost mandatory. This leads to an important question: Is a risky action really voluntary if there are few options for acting otherwise?

In this country, for example, no one points a gun at anyone's head and orders that person to take up a certain profession. At the same time, it has long been accepted that certain occupations carry inordinate risks. Most coal miners know of the risks that come with their line of work, yet they continue to mine coal. In times of high unemployment, unskilled workers rarely turn down jobs in dirty chemical plants.

In fact, occupational hazards have always been more acceptable to people than nonoccupational ones—a double standard reflected in laws that permit workers much greater exposure to chemicals and radiation than the public. Many of the residents of the Love Canal neighborhood in Niagara Falls, New York, who protested most strenuously about the toxic wastes in their backyards worked in local chemical plants where they were exposed to unhealthy fumes every day.

A chief reason for this double standard is that workers believe that risk exposure is what they're being paid for. (And some are very well paid indeed.) And since the consequences of this risk exposure often don't surface for many years, the risk doesn't stir up a sense of dread.

Furthermore, in some regions, certain careers are pretty much preordained. A steel worker in the Midwest, a coal miner in Appalachia, and an auto worker in Detroit have few job choices. It's work in the plant, mill, or mine, or not at all. This absence of choice is dramatically underscored every time the smoke stacks stand idle. Their social, emotional, and familial roots in industrial communities doom workers to occupational risk exposure. They know what the risks are, but take them in order to survive-in-place. "You never balance the wage against the risk," explained one worker, interviewed by Dorothy Nelkin and Michael S. Brown for their recent book, *Workers at Risk: Voices from the Work Place*, "you balance the wage against the alternative. And the alternative is starving."

Delayed consequences, lack of alternatives, and the necessary peril of some types of high-risk work so as to benefit a larger group ("someone has to do it") don't apply in every case. There *are* many risks that are chosen quite freely, without the duress of economic necessity and other extenuating circumstances. These are risks no one has to take.

In recent years, for example, a growing number of Americans—"conquistadors of the useless," in one adventurer's words—have taken up hang gliding, sky diving, hot-air ballooning, mountain climbing, white-water canoeing, scuba diving, and other daredevil sports. Few psychiatrists attribute this risky behavior to some kind of death-wish; the reasons seem to be more mundane. Eric Rosenfeld, a Manhattan attorney who has been climbing mountains for more than two decades, calls this activity "quite addictive" in terms of its risk factor. "It's something I kind of need as a counterpoint for the office situation, which can be quite dull," he told a reporter, adding that he has "a built-in denial of risk that's called surviving. I have an intellectual appreciation that it's risky," Rosenfeld said, "but I sit in my law office and tell myself that after twenty years of climbing I'm still here. It's always the guy on your right or on your left, but it's never you."

Some psychiatrists feel that most men take part in these sports to assert their masculinity, which in recent years has been curbed in many ways. This doesn't explain, however, why there has been a sudden surge in women participating in risky sports.

Dr. David Klein, a social scientist who has studied risk-taking at great length, predicts that while blue-collar workers used to be the "thrill-seekers," the ranks of executive daredevils will swell as white-collar jobs become more routine. "The basis for all this," Klein told *The New York Times*, "is the enormous emphasis our society places on performance and high virtuosity in that performance. If it cannot be performed on the job it is going to spill over to recreation." As Carl Boenish, a forty-two-year-old Californian who has parachuted off buildings and bridges, puts it: "Jumping is like a knife cutting through the malarkey of life."

Risky behavior that is entirely voluntary can also be propelled by vanity. Thousands, and perhaps millions of people every year, nagged by illness or injury, refuse to go to the doctor, fearing that they will be told they need to have surgery. Yet many others *elect* to have surgery that is unnecessary and which most people would consider frivolous.

One of the most common elective surgical operations, the facelift, is well accepted; some celebrities joke about their new faces on the *Tonight* show. Less publicized is the remarkable number of "breast-lifts" done on American women. Over one million women, according to the American Society of Plastic and Reconstructive Surgeons (ASPRS), have already had their breasts enlarged with plastic implants—a number that breaks down to one in every one hundred women over the age of eighteen—and another seventy thousand will probably join them this year. All of this despite the fact that a Food and Drug Administration (FDA) scientific advisory panel has urged the FDA to investigate evidence that silicone-gel-filled devices used for breast enlargement may be unsafe. According to one ASPRS member, Dr. Norman Huge, complications can be expected with 25 to 33 percent of the implants done with these devices. Yet, in a statement to the FDA, the ASPRS has termed small breasts "underdeveloped" and called them "deformities" that are "really a disease."

Ambition is another factor that causes people to open their

arms to risk. Many people—from high-powered business executives to baseball managers—choose to pursue stressful careers despite the many hazards (ulcers, hypertension, and broken marriages, to name only three) associated with them. They do so, it has always been thought, because the rewards that accompany such careers are often great: high salaries, prestige, fame, power.

Some recent studies, however, have found that some people who lead the stressful life are not so much hooked on the rewards as on the stress itself. They get pleasure from personal strife. Referring to the so-called "Type-A personality" who often engages in a continuous struggle with other people, and against circumstances within (and beyond) his control, Dr. Paul J. Rosch, president of the American Institute of Stress, has said that this individual "perhaps becomes addicted to his own adrenaline and unconsciously seeks ways to get those little surges." Adds Dr. Joel Elkes, emeritus professor of psychiatry at the Johns Hopkins Medical School: "Risk-taking and extreme stress produce a pleasurable arousal, followed by a feeling of release."

Most chosen risks, however, are neither purely voluntary (like hang-gliding) nor accepted with little alternative (like working in a coal mine). These are the risk choices that keep people awake at nights, locked in indecision, and that provide millions of jobs for government regulators, scientists, and researchers. These risks may not be great, or they may be associated with such wonderful benefits that the danger seems bearable. For most people it's not a matter, for example, of eschewing all salt, but of cutting down. We drive but we drive "safely"; we drink "in moderation." Millions of people live, quite happily, on flood plains or in earthquake country. There is little black and white in this twilight zone. It is where individual values, judgments, and priorities—all affected by knowledge, or the lack of it—come into play.

Faced with difficult risk judgments, we commonly make decisions in one of three ways: the technical approach, in which we consult all the latest statistics, odds, and official findings about a purported risk; the contrary approach, in which, rather than taking numbers into account, we base our action on what is known as "revealed preferences" (how most people have most often acted in the past reveals an "acceptable" activity); and the third and perhaps most common approach, in which we try to take *everything* into account, from the latest studies to a recent story in the newspaper to a personal brush with the same risk in the past (this might be called "empiricism").

All of these methods have their shortcomings, of course. It has been said, for instance, that the technical approach relies on the "countable" instead of what counts. And the one inevitable thing about the trial-and-error approach is that there *will* be errors.

In most cases, however, when it comes to making a decision about risks, we "muddle through," as Charles Lindblom put it. We juggle a number of factors—consciously and unconsciously. We conduct our own cost-benefit analysis. (Almost every risk is a calculated risk.) We ask ourselves: Are the possible ill-effects of the action felt immediately or delayed? Is the risk known with certainty or not? Is exposure to the risk essential or a luxury? Are there any alternatives to taking this risky step? Is it an occupational hazard? Are the consequences reversible or irreversible? And less frequently: Will my chosen risk impose a risk on someone else?

It is in these ruminations that knowledge becomes so important. Facts and information often make decisions easier, and generally improve the chances that the decision will end up being a wise one. Life, unfair to begin with, becomes even more imbalanced when it is lived in the dark. But, as we will see in this book, information is not an end in itself. Sometimes it is not even a helpful tool. Information about risks is often inadequate, overly complicated, or just plain wrong. In some cases, raw data are twisted for self-serving ends; in others, the public has difficulty assimilating and understanding truly useful information. And even when the information is universally endorsed by experts (which it rarely is), many people still arrive at erroneous judgments about it: they use adaptive strategies for reducing the information to simple terms they can understand, and in so doing distort it. Hang-gliders point out, for example, that only about eight in every ten thousand participants in this sport die every year in accidents. This seems like a low figure, but it is in fact the highest "kill-ratio" in all of sports; if everyone in America got into hang-gliding, the fatalities associated with that sport would jump from about thirty per year to over 150,000.

For most people, immediate gratification overwhelms the possibility of risk. In the early 1980s, 160 people tried to purchase and move into attractively low-cost homes vacated by families living around Love Canal. When asked why she would want to live in such a place, one would-be buyer replied, "The whole world's a chemical dump, so what's the difference?" Another potential buyer rationalized that he knew people who used to live at Love Canal who had no medical problems and people who live elsewhere who do.

Few would agree completely with the reasoning put forth by the first person and clearly the opinion of the second was based on limited personal observations. But the major impetus for the choice to buy a house at Love Canal was economic. Would-be buyers realized that houses in so undesirable an area must be sold cheap, providing an opportunity to purchase a home immediately they otherwise might have to work twenty years for. So powerful a force was that perceived benefit that they blinded themselves to the clearly documented Love Canal risks. As Benjamin Franklin phrased it: hunger never sees bad bread.

Widely circulated statistics on lung cancer can't conquer the need for a cigarette, and the latest reports linking cholesterol to heart disease have a hard time standing up to a bowl of chocolate-chip ice cream. Hard data are at a further disadvantage because of public awareness that scientists and other experts involved in risk assessment are sometimes wrong; scientists may be precise, but they are fallible. Furthermore, whether real or contrived, scientific controversy confuses many people and often leads them into choosing a risky habit, harmful food, or dangerous drug. Some industries, such as the giant tobacco, dairy, meat, and egg consortiums, encourage scientific controversy through the help of paid scientific allies, just to keep dispute simmering over an otherwise sound scientific consensus. Then, when making a choice, many persons weigh the faulty scientific data as equal to that coming from the mainstream of science.

In making choices about risky practices or products, we weigh the evidence, temper it with guesses about the risk probability, availability of alternatives, costs, and benefits, and apply our personal values. In so doing, we often end up making the same choices as someone who blunders into a decision in blissful ignorance.

Thus, while chosen risks come about through what we call free choice, such choices are obviously influenced by a host of factors, subtle and explicit. These factors interact in a decision-making process both conscious and subconscious, resulting in what we perceive as our free choice.

Close-Up (1)
SMOKING

"Mr. President, I want to make it clear I am a tobacco user. It is a curse. It is the worst curse that has been inflicted on me that I know of. I got into it at age seventeen and I have been puffing two packs of these things a day since age seventeen, thirty-five years ago. I know it is going to shorten my life. I know it will probably kill me. I despair the day I ever got addicted to this horrible mess. Thank God my children are not addicted. I am not proud of the fact that I do smoke. I wish I had had the courage to quit a long time ago . . ."

—Senator Thomas Eagleton, during a
Senate debate over continuation of
tobacco price supports, July 14, 1982

"Why do I smoke? We're all going to die in a nuclear war anyway."

—John Demeter, a magazine editor
in Cambridge, Massachusetts

Suzanne Rosenberg has been smoking a pack of cigarettes a day, on the average, ever since she left home in Los Gatos, California, to go to college fourteen years ago. During this period she has made only one sustained effort to break the habit, and that lasted only four months.

It's not that she needs to be shown that smoking is bad for her. "I recognize the risks," she says. "I've read all the shocking reports—I'm convinced!—and I know I'm not immune." Already, at the age of thirty-one, she notices that smoking has given her a cough and that it cuts her stamina. She feels guilty because she knows her habit adversely affects her husband (who suffers from allergies). Smoking in public makes her feel self-conscious "all the time"; she calls this the "smoking-pariah syndrome."

Yet she continues to smoke. She explains that she does so because she finds smoking orally gratifying and pleasurable. She likes the taste of tobacco smoke. A high-strung person, she uses cigarettes to relax. "When you're nervous," she says, "it feels good to have something in your hands." She finds smoking both a "social thing" and "a shield in social situations." Today, Rosenberg, a political science instructor and television production assistant from Englewood, New Jersey, has vague hopes—but no firm plans—for kicking the habit. She promises to stop smoking when she gets pregnant, but until then it's quite unlikely that she will quit. The reason? "Willful disobedience," she says. "A crisis of will."

It has been more than twenty years since the Surgeon General identified smoking as the nation's most significant source of preventable illness and premature mortality. Today, the annual death toll—perhaps three hundred fifty thousand people—exceeds the number of American lives lost in battle in World War II. Yet over 30 percent of American adults (and 20 percent of youths aged twelve to seventeen) smoke. For them, as for Suzanne Rosenberg, smoking is an acceptable risk.

Over fifty million Americans smoke and impose their chosen risk on millions of others who involuntarily inhale the toxic components of burning cigarettes. The medical, scientific, and statistical evidence that smoking is bad for health is overwhelming and convincing beyond the shadow of a doubt—except to industry trade groups, such as the Tobacco Institute, which have a vested interest in seeing that people continue to smoke. The smoking gun in this medical mystery is—smoking.

Why do so many Americans choose to expose themselves to this rather serious risk to their health? Are they misinformed or uninformed about the risks, oblivious to the warnings, or simply seduced by slick advertising?

One thing is certain: It is a myth that, in regard to deleterious health effects, smokers have "heard it all before." A Federal Trade Commission report noted not long ago that about half of all smokers have still not accepted the proposition that "cigarette smoking is dangerous to your health" and can cause lung cancer. The report stated that there is even less general acceptance of smoking's link to specific health consequences such as emphysema and heart disease. Seen in this light, R. J. Reynolds' controversial advertising campaign, which started in early 1984 and called for an "open debate" on the unsubstantiated perils of smoking, was far shrewder than many gave it credit for.

The individual decision to smoke and to continue doing so arises from a web of personality traits, perceived needs, and external influences.

According to government data, adult smokers, when compared with nonsmokers, tend to be "risk-takers" who are extroverted, compulsive, defiant. They are more likely to be divorced or separated. They consume more alcohol and coffee than nonsmokers; they take more psychoactive drugs. Housewives are more likely to smoke than working women; blue-collar male workers smoke more heavily than their white-collar counterparts. Among teenagers, those who smoke are more likely to have family and friends who smoke; they are more outgoing and rebellious. (Thirty percent of the smoking teenage girls said they had had sexual relations, versus 8 percent of the nonsmokers.)

Extensive research has found roughly six distinct reasons why people smoke. Some people are stimulated by cigarettes, and feel that they help them "get through the day." For many, cigarettes act as a tranquilizing agent, a crutch that helps them handle personal problems. Others find satisfaction in simply handling cigarettes because they have a strong need to keep their hands busy. About two-thirds get real "pleasure" out of smoking; they find that smoking reduces their negative feelings and makes them "feel good." There are also those who are psychologically addicted to smoking, and finally a number of people who simply smoke out of physical habit. (Among young people, peer pressure is a strong force promoting smoking, since many want to be "one of the crowd.") A person may smoke for any one or a combination of these reasons. Studies show that women tend to smoke mainly to reduce anxiety, while many more men smoke purely for "pleasure."

Those who are stimulated by cigarettes or simply like handling them find it relatively easy to stop smoking. So do those who get pleasure from cigarettes, provided they have carefully thought about the harmful effects of smoking. People with a craving or psychological addiction to cigarettes and those who use them as a crutch have a hard time quitting, as do habitual smokers.

Nonetheless, as the Surgeon General's 1982 report on smoking and health pointed out, people can quit if they try hard enough. Ninety-five percent of those who have successfully quit have done so without organized programs. And it appears that stopping "cold turkey" is more effective than cutting down gradually.

Most people quit either because they are suffering from some serious consequence of smoking, or because they have finally responded to the medical warnings. It has been found that brief and simple advice from a doctor is potentially the most cost-effective way of getting people to quit. On the other hand, some studies indicate that human decisions are not focused on final outcomes (in this case lung cancer, emphysema, and so forth) but rather on incremental stages. People move ahead one step at a time, regarding what lies several steps ahead as irrelevant. Thus, stressing the long-term consequences of smoking to young people is not very effective, because these consequences seem too remote; most young people who smoke know many others who smoke, while few know anyone suffering from lung cancer. Lung cancer, emphysema, and the other serious diseases caused by smoking look like low probabilities from where they stand: on the adventuresome side of forty.

Other studies have shown that adults habitually tolerate risks even if they mean a great potential loss—provided that the *probability* of that loss is low. Thus, the health consequences of smoking, taken on their own, are seen as serious but of low probability, and many people are willing to take the risk.

Contributing to the problem is the Tobacco Institute's seductive and reassuring advertising, which effectively dilutes the warnings issued by the Surgeon General and others. The Institute has regularly been able to muster some scientific support to its side, successfully conveying the erroneous impression that scientists are evenly split on the subject of smoking and its dangers. This, in turn, lulls many smokers into believing that the final word isn't in yet on the link between smoking and lung cancer, or some other risk. Then, too, many smokers believe that their choice is between cigarettes and frayed nerves—or dying at age eighty-one instead of eighty-two—and so they continue to smoke.

Moreover, although some smokers have switched to filtered cigarettes or to brands low in tar and nicotine as the medical evidence against smoking has mounted, this is often only a halfway measure at best.

The tar and nicotine content of cigarettes has long been a focus of concern for medical researchers. To combat bad publicity, cigarette manufacturers reduced the tar and nicotine content of cigarettes by an average of almost 50 percent between 1955 and 1975. Today, more than 50 percent of the cigarettes sold in the United States have a reduced tar and nicotine content.

This reduction seems to reduce lung cancer rates slightly but its overall health effects are unproven. A panel of National Academy of Science medical experts found that people who smoke low tar and nicotine cigarettes often compensate for the reduced levels of these ingredients (which they crave) by increasing their consumption, thus canceling out whatever benefits low tar and nicotine cigarettes provide. Another study found that half of those who smoke low tar cigarettes admit to blocking the filter vents which are supposed to dilute the smoke.

So milder cigarettes are only a partial answer. As the Academy's panel suggested: "Smokers who want to reduce the hazards from their cigarettes are well advised to quit smoking entirely."

Even the growing interest in nontobacco cigarettes—some tobacco companies are reportedly exploring this option—may not prove fruitful. Tobacco substitutes have never caught on, going back at least as far as 1839, when a U.S. patent was taken out on a sunflower-leaves-and-rhubarb cigarette. Smokers want their nicotine. And even nontobacco cigarettes produce smoke, and where there's smoke there's carbon monoxide.

How dangerous is cigarette smoking? In 1982 about 111,000 people died of lung cancer, and an estimated 85 percent of these deaths were directly due to smoking. Lung cancer killed 80,000 men and 31,000 women, reflecting lower cigarette use among women. But women have been smoking cigarettes in increasing numbers—they have, indeed, "come a long way." And, as former Secretary of Health, Education, and Welfare Joseph Califano pointed out: "Women who smoke like men will die like men." In 1984 the lung cancer death rate for women was expected to surpass that for breast cancer, the leading cause of cancer death among women. And according to the American Cancer Society, less than 10 percent of people with lung cancer survive five years.

Based on extensive scientific data, we know that those who smoke more than a pack a day have three to four times the general cancer death rate of nonsmokers. For smokers, the lung cancer risk is eight to fifteen times that for nonsmokers.

Seen in another way, the statistics tell us that a twenty-five-year-old man who smokes twenty cigarettes a day cuts his lifespan by 4.6 years. If he smokes forty or more cigarettes a day his lifespan is reduced by 8.3 years. It is estimated that every cigarette smoked reduces the lifespan by five and a half minutes—just slightly less than the time spent smoking it.

In February 1982, Surgeon General C. Everett Koop, making an annual report on smoking and health, went further in linking cancers other than lung cancer to smoking. He reported that cigarette smoking is now believed to be a contributing factor in cancer of the urinary bladder, pancreas, and kidney, and that it may be associated with stomach cancer. Many medical scientists feel that the ultimate cause and effect relationships between these cancers and cigarette smoking will eventually be demonstrated.

The Tobacco Institute's response to these statistics and to the carefully designed studies upon which they are based was consistent with its past positions. The Institute maintained that whether smoking causes cancer is still an open question and stated that the Surgeon General's latest conclusions were not new. This implied that there was nothing urgent or alarming in the Surgeon General's report and hence no cause for the government to take stronger antismoking action—or for smokers to stop smoking.

Cancer is not the only increased risk associated with smoking. Smokers have 1.5 to 2.0 times the risk of dying of coronary artery disease than nonsmokers. (Smokers who stop immediately reduce their coronary risk by a half.) According to government research, cigarette smoking is responsible for up to 30 percent of all heart disease deaths in the United States—about 170,000 a year. "Unless smoking habits of the American population change," Everett Koop said, "perhaps 10 percent of all persons now alive may die prematurely of heart disease attributable to their smoking behavior."

Massive public health education, restrictions on cigarette advertising, and limited bans on smoking in public have reduced the proportion of Americans who smoke. One study shows that 55 percent of all men smoked in 1955, 37 percent today. After a great spurt, the number of female smokers has leveled off at around 30 percent. Another report revealed that 29 percent of teenagers smoked four years ago, 20 percent now. Per capita consumption has fallen to 3,512 cigarettes a year, the lowest number since World War II.

But the medical and scientific testimony against smoking has been blaring for more than two decades. Why have over thirty million Americans stopped smoking only recently? Why do 90 percent of all smokers say they would like to quit?

Certainly, the physical fitness craze has had a great deal to do with it. New quit-smoking programs have proven increasingly effective (although up to 75 percent of those who quit eventually suffer a relapse). But perhaps most significantly, smoking's social standing has suffered; in some circles it is no longer even acceptable to smoke. And among young people it may no longer be "cool." A University of Wisconsin study released in early 1984 showed that people pictured with cigarettes were ranked by college students as being less sexy, less honest, and less mature than the same people pictured without cigarettes. "My guess is that our culture is changing," said Marshall Dermer, an associate professor of psychology who conducted the study. "Older people have not been exposed so strongly to negative aspects of smoking."

Nevertheless, according to most accounts, the number of smokers in the United States is rising by about 1 percent a year (the same rate applies worldwide) because of increasing population and the steady number of women who smoke. Among smokers the proportion of *heavy* smokers is actually increasing. In 1982, 636 billion cigarettes were sold in the United States.

A major reason for the continued good health of the cigarette industry is that the six major American tobacco companies still control the public debate. They spend about $1.5 billion annually in advertising and promotion, the largest amount spent on any product in this country. (By comparison, the government's Office on Smoking and Health has a yearly budget of about $3.5 *million*.)

Besides increased sales, these expenditures encourage, in the opinion of some observers, relative silence from the print media on the subject of smoking. *Time* and *Newsweek*, for example, receive in excess of $30 million a year in cigarette advertising; neither newsmagazine has printed a cover story on the hazards of smoking since 1970, while devoting covers to AIDS, Legionnaire's disease, and toxic shock syndrome. (The June 6, 1983, issue of *Newsweek* contained one story on nonsmokers' rights—but no cigarette advertising. The tobacco companies had not pulled the ads; *Newsweek* did.)

Cigarette commercials have not appeared on television since 1970. This turn of events came after John Banzhaf, director of a public interest group, petitioned the Federal Communications Commission to use the Fairness Doctrine to require television and radio stations to give "equal time" to antismoking ads. The antismoking spots subsequently aired were so strong that the cigarette industry decided to voluntarily remove their ads from TV and radio (resulting in the end of the antismoking messages as well).

But this was not a death blow to the cigarette industry; it was, in many respects, a bonanza. Cigarette companies switched to less costly advertising in magazines and newspapers, avoiding the threat of being met with "equal time" ads. With the money they saved on advertising, they launched a two-pronged promotional effort consisting of corporate philanthropy to cultural and sports organizations, and a massive print advertising push.

Suddenly there was a Virginia Slims Tour (tennis), the Marlboro Cup (horse racing), Kent Ladies Golf Classics, the Winston Team America Series (soccer), the Marlboro British Grand Prix (car racing), and "Camel Scoreboards" in the sports sections of one hundred newspapers. The cigarette manufacturers also courted respectability by sponsoring such cultural events as the Salem Spirit Concert series, the New York Kool Jazz Festival, and many art exhibitions. (Philip Morris sponsored the Vatican exhibit at the Metropolitan Museum of Art in 1983.)

At the same time, many of the new cigarette ads in magazines and newspapers, and on massive billboards, began emphasizing natural beauty—the crashing of waves, snow-capped mountains, and rolling plains. The models who appeared in them seemed to grow increasingly younger and attractive; smoking was "sexy." In all of the ads the smoker was shown in control— a subtle but powerful message to the millions of smokers who light up to reduce tension. These ads appealed to young people and those who "think young," and no wonder: The land of cigarette advertising is a dream world bereft of coughing, dirty ashtrays, or cancer.

The Surgeon General's warning is included in all of these ads, not in matching color and bold print, but in unobtrusive black and white that detracts little from the ads' impact. "Ultra low tar" and filters are promised along with satisfying taste, implying safety and reward all rolled into one cigarette. None of the ads indicate, of course, what the beautiful people in them would probably look and feel like twenty years hence if they were to smoke heavily. The mountain vistas and seascapes are silent about what has happened to those who long ago heeded the advice, "What are you waiting for? Come and try one." No cigarette ads show smokers with emphysema gasping for breath, or those with lung cancer dying in their hospital beds. Yet these visual images are more relevant to most cigarette smokers than rolling waves and rugged mountains.

Ironically, many ads are aimed at the sports- and fitness-minded. Going after the youth, female, and health-conscious market in one shot, Virginia Slims put out a full-page offer for a warm-up suit called the Ginny Jogger—perhaps the only fitness ad ever to appear with a health warning from the Surgeon General printed at its bottom.

One of the most effective antismoking devices making the rounds in this country tackles one of the industry's advertising themes head-on. The documentary *Death in the West,* made for Thames Television in Great Britain several years ago, has been shown to children in forty states. It includes interviews with six cowboys who are dying from diseases caused by smoking. One scene shows a man astride his horse in the distance, with the Marlboro theme on the soundtrack; a close-up shot reveals that he has an oxygen tank strapped to his saddle and tubes sticking out his nose.

But, in the final analysis, despite such antismoking propaganda, millions of people choose to smoke. No one forces them to do it.

Smoking is a chosen risk. Many people who smoke recognize the risk and respond to those who complain about smoking: "It's my problem. It's none of your business."

But this is not so. What is a voluntary risk for one person is an imposed risk on another.

One form of this imposed risk is through involuntary or "passive" smoking—nonsmokers inhaling smoke generated by others. In the 1982 Surgeon General's report, Dr. Edward N. Brandt, Jr., an assistant secretary of Health and Human Services, said, "Prudence dictates that nonsmokers avoid exposure to secondhand tobacco smoke to the extent possible." Dr. Brandt's comment was made in response to evidence produced in three studies that showed that the wives of heavy smokers had a higher risk of developing lung cancer than the wives of nonsmokers.

Passive exposure to cigarette smoke is hazardous because the carcinogenic and toxic chemicals in the cigarette are released into the air unfiltered. One study indicated that a nonsmoker sitting at a crowded, smoke-filled bar inhales in one hour approximately the same amount of dimethylnitrosamine, a carcinogen, as that inhaled by someone who smokes about twenty-five filtered cigarettes.

The tobacco industry has fought every effort that would require them to identify publicly the chemicals added to cigarettes, and for good reason: some four thousand known compounds are generated by burning cigarettes, including many carcinogens and such substances as vinyl chloride, arsenic, and insecticide residues. All of these, of course, are passed on to bystanders. Radioactive substances (polonium-210 and lead-210) are also found in cigarette smoke. In a study published in the *New England Journal of Medicine* in 1982 two researchers stated that a one-and-a-half-pack-a-day smoker may be getting the equivalent of three hundred chest X rays annually. Another study found that only 7 percent of the radioactivity in tobacco is absorbed by the smoker himself—some 30 percent is transferred laterally in the so-called "sidestream" smoke to his companions.

During an interview in 1983 Surgeon General Koop declared that "we are going to prove without any doubt how dangerous" the problem of "second-hand" smoking is to the nonsmoker; recent studies, he said, "showed beyond any shadow of a doubt sidestream smoking is very dangerous to your health." Koop suggested that the warning on a cigarette pack might be more effective if it read: "The smoke from the cigarette you are about to exhale is dangerous to the health of your children." (The Tobacco Institute asserts that the preponderance of scientific testimony does not show that ambient tobacco smoke is harmful.)

Because they consider cigarette smoke annoying—not necessarily harmful—nonsmokers have started complaining loudly. A Roper survey found, in 1978, that the percentage of people favoring separate smoking and nonsmoking sections ranged from 61 percent (for offices) to 91 percent (trains, planes, and buses). Roper called the nonsmokers rights movement "the most dangerous development to the viability of the tobacco industry that has yet occurred." (This is shown by the following statistic: If each American smoker smoked just one less cigarette each day, the tobacco industry would lose about half a billion dollars a year.)

Antismoking sentiment is growing more "dangerous" by the year. In 1984 Everett Koop startled some observers by calling "a smoke-free society by the year 2000" this country's top health goal. Representatives from the American Cancer Society, American Lung Association, and American Heart Association promptly formed a coalition to take up this challenge. The Tobacco Institute condemned this effort, stating: "The new prohibitionists have no place among a free people."

Aggressive antismoking sentiments inspired a host of new state and city laws, the most comprehensive of which was the state of Minnesota's Clean Indoor Air Act, which took the radical step of prohibiting smoking everywhere unless it was specifically permitted. Among other things, the law requires No Smoking sections in restaurants. A Minnesota legislator, Phyllis L. Kahn, reports that smoking has declined in the state because it is simply more difficult to smoke. "Few people continue to smoke," she says, "when it is pointed out to them that it is against the law." A 1980 survey found that 92 percent of nonsmokers in Minnesota favored the law—and, amazingly, so did 87 percent of heavy smokers. Some antismoking activists claim that most smokers are actually grateful for the new laws because the only way they might quit smoking is if it's not allowed. But Charlie Fahvort, a businessman in Washington, D.C., interviewed by The New York Times, expressed another view: "Nobody knows better than me that smoking is bad for me. But people who do not smoke are a little too arrogant."

Voters in San Francisco approved in 1983 a measure that required businesses to provide agreeable working conditions for nonsmokers. (The tobacco industry spent $1.3 million trying to defeat the measure.) In a landmark decision a U.S. Circuit Court of Appeals ruled that nonsmoking federal government workers who are hypersensitive to smoke are eligible for disability unless they are transferred to a job in a "safe environment."

Some employers have found that favoring nonsmoking employees helps keep costs down. One study found that the average worker who smokes uses up about 50 percent more sick leave than one who does not smoke. The Wall Street Journal recently reported that managers are increasingly requesting employment agencies to send them people who do not smoke. Columnist William Safire has noted, only somewhat in jest, that smokers now comprise a new "oppressed minority group."

While people can choose to smoke or not—and to leave rooms or restaurants to avoid cigarette smoke—children and the unborn are utterly defenseless. Many studies have found a decrease in birthweight for infants born to mothers who smoked during the second half of pregnancy. Other studies show that children whose mothers smoke may suffer reduced lung development. These findings provoked former Surgeon General Dr. Luther Terry to write: "Cigarettes are *child abuse*."

Another way people who choose to smoke strongly affect nonsmokers is reflected in the following statistics: estimates of the cost of smoking to society as a whole in the United States range from $35 to $49 billion (this includes direct health-care costs, lost wages, and decreased productivity). These figures do not take into account what Everett Koop calls "something else far worse—the cost of human dignity and quality of life of the families left behind by people who die due to smoking."

By one estimate, every nonsmoking adult in this country pays at least one hundred dollars a year in taxes and increased insurance premiums to help cover the health-care costs piled up by smokers. Insurance companies have long been criticized for not offering lower life insurance premiums to nonsmokers, who tend to live several years longer than heavy smokers. (Some companies, including Prudential, do offer discounts, and this trend is beginning to extend to health, fire, and even auto insurance.)

Smoking, a risk which appears so personal on its face, imperils others in at least one other extremely destructive way: Cigarettes are the leading cause of deaths due to fires. About 65,000 fires every year—around 8 percent of all residential fires in the United States and about a quarter of all residential fire-related deaths and injuries—are smoking related. This translates into 2,300 deaths and 5,800 injuries annually and a property loss of $300 million. (One study showed that 39 percent of the victims of these fires were "innocent victims.")

Cigarettes cause fires because, when not being puffed, they can burn for twenty to forty-five minutes. A cigarette dropped on upholstered furniture will usually smolder, then flame within fifteen minutes. The devastation of cigarette-caused residential fires is dramatically depicted in a recent public service message made for television by the New York City Fire Department. The film shows Betty Brinkley, a Camden, New Jersey, housewife, who was badly disfigured in a 1977 fire that killed her husband and three children. Brinkley's husband fell asleep while smoking a cigarette.

Yet even tragedies such as this have not inspired a solution: legislation that would mandate self-extinguishing cigarettes. The Tobacco Institute has staunchly opposed these proposed laws, and legislators in several states, including New York and Connecticut, have killed such bills.

The tobacco industry opposes these measures because they would require the use of cigarette papers that would allegedly adversely affect the taste of cigarettes. Nat Walker, a spokesman for R. J. Reynolds, says that "by the time you have the kind of paper that leads to self-extinguishing, the taste is not good. Good taste for smokers is extremely important. If consumers don't buy our products, we're out of business."

William D. Toohey, Jr., director of media relations for the Tobacco Institute, contends that "the technology that would cause cigarettes to 'self-extinguish,' or go out when not being puffed, would inevitably raise levels of tar and nicotine, clearly not what consumers want these days," and adds, "If consumers were denied what they want through legitimate retail sales, bootleggers undoubtedly would be glad to provide them cigarettes of their choice and avoid the applicable state and city taxes."

Proponents of self-extinguishing cigarettes, however, do not agree that it is "inevitable" that these products would taste worse or contain more tar and nicotine. They point out that cigarettes are already chemically altered—to provide a long, slow burn. And they say that at least two brands sold in the United States—More and Sherman's—have been shown to be more fire-safe than others.

The battle lines are now drawn between the tobacco industry and those who want to reduce the risks of cigarette-caused fires. The industry is certain to put up a long and hard fight to keep cigarettes burning like fuses, while remaining generous with donations to general fire prevention efforts. (The Tobacco Institute recently donated $640,000 to the National Fire Protection Association's smoke detector program.) The American

Medical Association and the American Burn Association have endorsed efforts to make a safer cigarette but it is unlikely that the federal government will take a strong stand. The government has long played an inconsistent—some would say, hypocritical—role on the smoking issue ever since the Surgeon General's 1964 declaration that smoking was dangerous. On the one hand, the government has mandated warnings on cigarette packs to discourage consumption, and on the other it has given generous federal price supports to tobacco farmers.

In an effort to convince people that the health consequences of smoking are serious and real, health organizations and some members of Congress in 1984 pushed for more specific labels on cigarette packages, and suggested that they be changed regularly (or "rotated"). Many in the advertising field saw the present warning as suffering from "wear out," having become so familiar that it is often overlooked. (One study showed that only one in fifty smokers acknowledged reading it.) Yet the Reagan Administration gave little support to this proposal, and the Tobacco Institute was against it, with good reason: The proposed new labels were much more explicit about the health consequences of smoking.

However, during the spring and summer of 1984, a number of Congressional committees gave overwhelming support to strengthening the wording on cigarette packages. Aware that stiff legislation might be passed, the tobacco industry decided to negotiate behind the scenes with health groups and legislators proposing the new laws. (Deleted from the bill, and proposed warnings, were any references to "addiction.") The Comprehensive Smoking Education Act was finally approved overwhelmingly by the House of Representatives on September 10, 1984. Representative Albert Gore of Tennessee, a major sponsor of the bill, said: "This has been a bitter pill . . . for [the tobacco industry] to swallow, but in so doing they have made stiffer punitive legislation less likely in the years ahead." The new warnings set to rotate every month read:

- Surgeon General's Warning:
  Smoking Causes Lung Cancer, Heart Disease, Emphysema, and May Complicate Pregnancy.

- Quitting Smoking Now Greatly Reduces Serious Risk to Your Health.
- Smoking by Pregnant Women May Result in Fetal Injury, Premature Birth, and Low Birth Weight.
- Cigarette Smoke Contains Carbon Monoxide.

The bill also required that warnings be 50 percent larger in advertisements than the old warning. In addition, cigarette manufacturers would have to disclose to the Secretary of Health and Human Services a list of the additives used in their products.

Surprisingly, some antismoking activists opposed the change in warning signals, on grounds that it would disarm what they feel is potentially the strongest weapon against cigarettes: product liability lawsuits. They feared that tougher warnings would only bolster the tobacco industry's main legal defense: that smokers voluntarily assume all risk by choosing to smoke. (No one has ever successfully sued a cigarette company.) They pointed out that the reason the tobacco industry was so adamant about removing any reference to "addiction" on the labels was that this would seriously weaken their "voluntary risk" defense.

The new warning labels still have to compete with billboards filled with beautiful landscapes and beautiful people, the implied safety of filtered and ultra-low-tar cigarettes, and promises of fresh tastes and satisfactions. Thus, smoking, although a personal decision, is clearly not an uninfluenced one. Those who smoke choose the risks not only from free volition, habit, and need, but also because they have been persuaded, influenced, propagandized, or confused, or because they remain oblivious or ignorant. The choice—while still their own—isn't as free as one might think, and it is a choice whose impact is not confined to the individual "freely" making it.

Close-Up (2)
SEAT BELTS

*"People just don't take driving seriously. They grow up with it
and take it as second nature. They never stop to think they're
driving three to four thousand pounds of steel down the road at
forty to fifty miles per hour."*
> —Frank Kramer, accident investigator
> for the California Highway Patrol

Americans have been ramming, battering, and killing each other
on the highways ever since that day in 1904, when—the story
goes—the only two automobiles in Kansas City, Missouri,
smashed into each other at an intersection. This year about
forty-five thousand Americans will lose their lives on the road,
and three million will be injured. Another three thousand will
be crippled for life.

As many as half of these deaths and injuries could be prevented if everyone wore seat belts. Nevertheless, a recent Gallup
Poll revealed that only 17 percent of the people surveyed reported wearing a seat belt the last time they got into their cars—
versus 28 percent a decade earlier.

National studies based on the *observed* use of seat belts have
established that in reality, only one in eight Americans—about
13 percent of the population—regularly buckles up. Whether
men or women, young or old, the figure is nearly the same.
(Drivers on the West Coast, for some reason, buckle up a little
more often.)

This eschewing of the seat belt is not caused by doubts about
its effectiveness: The Gallup Poll showed that the average American believes that 40 percent of those who die on the road would
be saved by seat belts. Most Americans recognize the dangers
in driving, and the usefulness of seat belts, and still refuse to
buckle up.

Why, despite years of regulation and multimillion-dollar advertising campaigns, do only a tiny proportion of Americans
wear safety belts? The chance of getting into an auto accident
is infinitely higher than being in a plane crash, yet nearly every
passenger on a jetliner buckles up before takeoff.

According to most surveys, the three leading explanations offered by drivers who don't wear seat belts are:

"It's inconvenient—too much of a hassle."
"I'm forgetful."
"They're uncomfortable."

Other popular excuses are:

"I don't need to—I'm really a good driver."
"I only need to wear it when I go on long trips or drive at high speeds."
"I don't want to be trapped in the car if there's an accident—it's better to be thrown free."
"I just don't believe an accident will happen to me. . . ."

Over the years, however, car manufacturers have made seat belts more convenient and comfortable. "Buckle-up" advertising campaigns have attempted to counter many myths about driving by stressing that:

- Good drivers are frequently victimized by mechanical failure or by poor drivers (as many as half of all serious accidents are caused by drunk drivers).
- 75 percent of all accidents occur less than twenty-five miles from home, and 80 percent of deaths and injuries occur in cars traveling under forty miles per hour.
- Unless you are fortunate enough to land on a mattress, being thrown from an automobile is twenty-five times more dangerous than remaining buckled in your seat.
- Everyone can expect to be in a crash once every ten years; for one in twenty it will be a serious accident.

Seat belts do make a difference. One study showed that those who wear seat belts have 86 percent fewer life-threatening injuries than those who don't. Yet seat-belt use is not catching on.

This has led some researchers and observers to suggest that many Americans actually enjoy the "kick" of taking their lives in their hands every time they hit the road. The automobile, Senator Daniel Patrick Moynihan has written, is "both a symbol of aggression and a vehicle thereof. . . . It is a prime agent of risk-taking in a society that still values risk-taking, but does not provide many outlets." This, says Moynihan, suggests "the seeming incompatibility of safe driving and mass driving."

Patricia Waller, an associate director at the University of North Carolina Highway Safety Research Center, points out that there are often four times more injuries each year from motor vehicle accidents nationally than the combined number of new cases of cancer, coronary heart disease, and stroke. "Yet how many people," she asks, "worry about protecting themselves from car accidents as much as from cigarette smoke?"

This risk-taking on a mass scale may be chosen, but it has horrible ramifications for others—including the widows, orphans, and friends of crash victims, and society as a whole. A 1981 study by the Insurance Institute for Highway Safety revealed that the annual cost to the American economy of motor-vehicle injuries was $20 billion. Only cancer causes steeper losses.

To wear or not to wear seat belts will remain a life-and-death issue in America even if inflatable air bags are eventually required in new cars. This requirement would not affect old cars, and safety experts strongly suggest that drivers and passengers wear seat belts even in cars equipped with air bags because of the threat of side- and rear-end collisions, which do not activate the air bags. A Michigan study of more than four thousand accidents revealed a further reason to wear seat belts: Occupant-to-occupant collisions inside the automobile caused or aggravated injuries in 22 percent of the crashes.

Highway safety has been a controversial subject ever since automobiles were introduced to America. But it wasn't until 1966, after traffic deaths first reached the fifty-thousand-per-year mark—and after much prodding from Ralph Nader and other consumer critics—that Congress decided fundamental automobile design played a major role in the highway death toll. In that year, it passed laws forcing the reluctant auto industry to set new crash-performance standards. As a result, today's windshields don't shatter as they once did, steering columns collapse on impact, and automobile interiors contain more padding. The government estimates that these changes have reduced the fatality rate by 30 percent, saving more than ninety thousand lives. It is estimated that today, a motorist can drive 1,600 miles with the same risk as someone who drove 1,000 miles in 1966. Current costs to the consumer for those six hundred safe miles: an estimated $370 extra per car (little more than the cost of many new-car "options").

The major contributor to this reduced risk was the requirement, starting in 1964, that every new American-made car be fitted with a lap safety belt. In 1968, shoulder harnesses were made mandatory, and in the same year the National Safety Council was granted about $50 million in free broadcast and print advertising for public service announcements encouraging seat belt use. "Buckle up for safety" and "Lock it to me" resounded over the airwaves.

It quickly became apparent, however, that most Americans would not heed this advice. Part of this was because the first belts on American cars were rather unwieldy. As early as 1969 Secretary of Transportation John Volpe was suggesting that the government had to go one step further in safety design and come up with an "automatic" answer to the crash-impact problem. Officials declared that Americans were just too lazy, forgetful, or dumb to take care of themselves. "The answer," Volpe said, "is the air bag."

Air bags, installed under the dashboard, would inflate in one-fiftieth of a second after a crash to provide a buffer between drivers and hard metal and glass. The insurance industry predicted that the bags—often referred to then as a "technological vaccine"—would save more than ten thousand lives, thousands of injuries, and billions of dollars in medical bills every year. But the auto industry was opposed to the costly deployment of these devices, and many consumers feared that the bag might fly up in their face, without warning, at any moment. It wasn't difficult for the auto industry to downplay, and eventually phase out, initial experiments with air bags. (For an account of the epic air bag battle, see Chapter 5.)

With the personal encouragement of President Nixon, who

said he would never trust an air bag, the Department of Transportation shifted gears in 1973. Rather than requiring air bags, it mandated that all new cars have ignition systems that would not work until the driver fastened his or her seat belt. This proved to be a colossal blunder. Many drivers were outraged when, even after buckling up, their cars would not start (a common design flaw). At the direction of Congress, the DOT quickly rescinded the rule.

The ignition-interlock system, as it was known, might not have been much of a help even if it had worked perfectly. One study showed that 40 percent of drivers had found a way to disconnect the system; almost 30 percent in one survey labeled the interlock system the "least-liked" feature of their new cars. Even those who allowed the system to work properly went back to their old pattern of (low) seat-belt use once the new law was negated.

Most auto makers retained a portion of the interlock system in subsequent models. When the key was put in the ignition, a buzzer would sound and a light flash, reminding drivers to buckle up. However, studies have shown that the drivers of cars that have this feature buckle up no more frequently than those in cars that do not nag, remind, or chastise. The Insurance Institute for Highway Safety calls the buzzer system a "public health failure."

On the other hand, a potentially significant "passive" safety system was developed by the Volkswagen Corporation. This was the so-called "automatic seat belt." When you open the car door the already attached shoulder harness swings out and around you as you take your seat. According to the Department of Transportation, people in VW Rabbits equipped with the automatic belt have 55 percent fewer fatalities than those in the same model car with standard belts.

Still, many drivers balk. Some don't like having no choice in the matter; others are afraid they would be locked in their cars when they would most want to get out (after an accident). One survey showed that 52 percent of those who generally fail to wear seat belts claimed that they would attempt to disconnect any automatic system. And so the passive belt has not caught on in Detroit, where many executives still believe in the old industry adage "safety doesn't sell" (or to put it another way, "safety must not be sold").

It's no wonder, then, that seat-belt use plummeted. Unfortunately, this came at a time when driving, for many people, was more dangerous than ever.

That's because Americans are driving subcompact cars in record numbers, and while small cars may save fuel, they cost lives. The fatality rate for drivers of subcompacts involved in accidents is twice the average rate. Also, as cars get smaller, more and more accidents involving "uneven matches" take place on the road. In a collision between a subcompact and a full-sized car, the people in the small car are about eight times more likely to be killed than the occupants of the big car.

This problem will soon grow much worse. With the decline of the "oil crisis," the sales of full-sized cars have shot up. Meanwhile, General Motors, Ford, and Japanese automakers are about to introduce, on a mass scale, subsubcompacts—or "minicars." Safety experts warn that highway fatalities may jump significantly by 1990.

In its early days, the Reagan Administration—in keeping with its regulatory cutback approach—determined that the best approach to auto safety had long been overlooked: cajoling.

"We can get quicker and bigger results by changing human behavior, not vehicle structure," commented Barry Felrice, an associate administrator of the National Highway Traffic Safety Administration (NHTSA). And Representative Robert J. Walker added that "there is too much focus on building cars safe for people unskilled in driving."

These sentiments echoed what auto makers had been saying for years. Roger Maugh, director of auto safety for Ford, says that his company believes "that the most effective strategy to reduce traffic fatalities and disabling injuries is to get motorists to use the seat belts that already are in their vehicles," adding that "It's unfortunate this country has invested more than $14 billion in seat belt systems [that] are used only 10 percent of the time."

The Reagan Administration would use motorists' poor safety habits against them: a "we-only-help-those-who-help-themselves" approach. In April 1981 the NHTSA proposed revoking, revising or delaying seventeen safety regulations or rulemaking procedures on a variety of items, from crash-resistant bumpers to air bags. This would save the ailing auto industry $500 million annually. Any gap in motorist protection, the NHTSA said, would be filled by a new $9.6 million publicity campaign to encourage seat-belt use. Undaunted by the general failure of past campaigns, NHTSA administrator Raymond Peck said the new campaign would work because it "applies quantum leaps forward in behavior research and motivation research." Peck declared: "We're going after behavior modification because it's cheap, and it can work."

Former NHTSA administrator Joan Claybrook, however, was unconvinced. She charged that "in order to gloss over its callous disregard for public safety, the administration needed to speak positively about highway safety. Thus, the safety belt promotion program was born." Claybrook pointed out that even if the unofficial goal of the program was met—to raise seat belt use by about 20 percent—it would "still leave 70 percent of all occupants unprotected."

One of the agency's early plans was to get the makers of Chinese fortune cookies to include such messages as "Confucius say man who wear seat belt save face" in their products. And with a $2,800 grant from the NHTSA, the National Safety Council sent a seat-belt safety kit to three thousand religious leaders. The kit included a spiritual message from Billy Graham ("O, Lord, Creator, leader of all mankind . . . let me find the daily strength to insist my fellow passengers protect themselves in safety belts when they are traveling in the car with me") and a responsive reading in which the congregation was instructed to repeat: "Help us to buckle up."

The bulk of the NHTSA safety belt campaign, however, was based on a .new "networking" plan which supplemented the usual public service announcements with personal appearances by "authority figures" in the community. Added to this were various incentives provided by major employers participating in this carrot-and-stick program, who offer prizes to workers who wear seat belts; in some cases, company representatives stand in the parking lot and pass out coupons for free meals at

McDonald's to drivers who come to work buckled up. The NHTSA also suggested that employers tell workers that they may lose 25 percent of their compensation benefits if they are injured in a car accident while not wearing a seat belt. (Some have suggested that Congress pass legislation that would allow insurance companies to refuse payment for injuries or deaths to those who fail to buckle up.)

General Motors embarked on its own incentive program in 1984: Henceforth, for one year after the purchase of any new GM car or truck, the owner and passengers will be insured by the auto company for $10,000 against fatal injuries, providing they are wearing seat belts at the time of the accident. In its advertisements GM declared that "seat belt user insurance could change the way people think about seat belts."

But many auto safety activists are cynical about the public relations approach. "We have a persistent myth in this country that if we were all perfect, cars would never crash," says William Haddon, Jr., a former NHTSA administrator and now head of the Insurance Institute for Highway Safety in Washington. "But the fact is that most deaths and injuries would not occur if people were properly packaged in cars." The heart of Haddon's argument (he favors the installation of air bags) is that machines are more malleable than people.

The dismal record of seat-belt publicity campaigns supports Haddon's views.

In 1977 the four domestic auto makers initiated a pilot program, a ten-week, $1.75-million media blitz in the Detroit area. A DOT report at the conclusion of this advertising campaign indicated that it got "no response" from the public; in one city, in fact, seat belt use had declined by 1 percent.

In the same year, the DOT reported that education and promotion efforts in the United States and abroad "have not been successful in increasing voluntary seat belt usage to an effectively high level." A 1980 National Academy of Sciences report said much the same thing. Many experts feel that even a wildly successful media campaign could not boost seat-belt use beyond the 30 percent mark. Well into its heralded "buckle-up" campaign, the Reagan Administration could claim only a momentary 4.4 percent increase in seat-belt use.

This has led many consumer activists to demand that the government take safety out of the hands of motorists and make automatic seat belts or air bags mandatory. Many of these activists cite other health and safety areas in which the government has had to step in to protect irresponsible consumers. "If pasteurizing milk was made optional again," points out Ben Kelley, a vice-president at the Insurance Institute, "some companies would be able to sell a half gallon of unpasteurized milk for twenty-five to thirty cents, and you'd find a lot of people buying it—and getting sick." In this view, giving people the option of buckling up is like selling unpasteurized milk and suggesting that everyone boil it before drinking it.

Even though the Insurance Institute thinks that media campaigns are, as Ben Kelley puts it, "undependable," it has produced several films itself to shock viewers into wearing seat belts. One, called *Faces in Crashes*, was sparked by Institute-funded research that discovered that the "trauma" of auto accidents was far worse than previously suspected. This study found that auto crashes in the United States cause about 114,000 severe facial lacerations and 25,000 severe facial fractures every year. Most of these injuries occur when heads smash into windshields, dashboards, or steering wheels.

The Institute argues that this study proves that air bags are necessary; it created *Faces in Crashes* to convince people that seat belts, too, are useful. This ten-minute short shows test cars smashing into walls; the dummies inside are showered with glass daggers. Real-life survivors of such crashes, bearing unsightly scars, testify in favor of seat belts. In the film, Institute chairman Bill Haddon points out that while wearing a seat belt is up to the individual, society as a whole pays. "These kinds of injuries in motor vehicle crashes," Haddon says, "tie up all sorts of medical personnel that could be better used for other things; tie up hospital beds and surgeons; and involve great expenses."

Ben Kelley feels the film was effective because a lot of people "fear scars and plastic surgery more than crippling injuries." One Institute study found that those who had a friend or relative injured in an automobile crash were more likely to use seat belts than those who had a friend or relative killed in an accident. This, said the study, indicated that "fear of being disfigured or disabled is more conscious and motivational in the use of safety belts than fear of death in a crash."

A new drive has been mounted to make the use of seat belts mandatory in the United States. In the past dozen years, laws that would require seat-belt use have been introduced in at least twenty-seven state legislatures, with only New York, in July 1984, enacting such a bill.

Studies of thirty-two countries—including Japan, Great Britain, France, and the Soviet Union—in which buckling up is mandatory have revealed that such laws provide some benefits, although these are by no means overwhelming. Seat-belt use is required in West Germany, but only about one-third of the drivers in that nation abide by it. In Ontario, Canada, fines for nonuse range from $20 to $100; compliance is estimated at 60 percent.

However, in most areas where mandatory seat-belt laws have gone into effect, the reduction in deaths and injuries has been disappointing. To cite one tragic example: French police revealed that Princess Grace of Monaco was not wearing a seat belt, as required, at the time of the 1982 automobile accident that took her life.

Several theories have been brought forth to explain why this is so. Among the hardcore who refuse to buckle up under any circumstances (an estimated 55 percent of American drivers) are a large number of those most likely to be involved in an accident—teenagers and drunk drivers. A Johns Hopkins University study showed that drivers who run red lights are less likely to wear seat belts than those who don't. A General Motors study revealed that drivers who tailgate other cars are less likely to buckle up. Other researchers feel that wearing a seat belt makes some drivers cocky—and so they subsequently drive less carefully.

Americans oppose a mandatory seat-belt law by about a two-to-one margin, polls reveal. Indeed, a survey conducted for General Motors showed that car owners, by a two-to-one margin, prefer the air bag to mandatory seat-belt use, even if the bag costs them about $300 to install in a new car. Americans, it appears, will pay almost anything to hold onto their right to be lazy.

At the same time, however, more than twenty states have passed laws requiring children under the age of four or five to ride in car seats equipped with special belts. (Approximately 90,000 children under the age of six are injured every year by motor vehicles.) In Tennessee—the first state to pass a child restraint law, in 1978—compliance has been estimated at only about 40 percent, yet infant auto deaths in the state dropped from seventeen in 1978 to five in 1982.

Lawmakers have decided that a child's safety is too important to be left to his or her parents. The parents, for now, are on their own.

Such contradictions abound in the field of auto safety. A 1982 Roper Poll, for example, showed that Americans opposed, by a 67 to 26 percent margin, the current drive to relax auto safety standards. Yet by an even greater margin they refuse to take one small, easy step toward safety by fastening their seat belts.

The question remains: Why, despite the statistics, the public service announcements, and the obvious advantages of seat-belt use, do so few Americans buckle up? After stripping away the excuses, many who have studied the problem believe that this risky behavior is primarily based on the feeling "it won't happen to me."

A national survey of motorists, by Peter Hart Associates, showed that 21 percent felt that "the chances of getting into an accident are so small that seat belts aren't really worth the inconvenience." Thirty-seven percent said "there's nothing that would make me use seat belts most of the time." In a 1979 survey for the DOT, the typical respondent guessed that the chances of being in a car crash over a year's time was one in one hundred. In fact, it's one in ten.

Nor, apparently, do all the pictures, films, and statistics that graphically show what happens to people in accidents have much impact on drivers who feel immune to this eventuality. The low level of seat-belt usage is consistent with studies of insurance-buying habits, which show that people generally opt for coverage on high-probability/low-damage risks rather than low-probability/high-damage ones. And the risk of having a fatal accident each time you go for a drive *is* relatively low, with only about one in every three and a half million automobile trips ending in death. "Repeated benign experience leads eventually to a habit of neglecting the belts," says Paul Slovic of Decision Research in Eugene, Oregon. "The extremely low probability of an accident on a single trip makes it unlikely that more than a few individuals will ever use seat belts voluntarily."

Nevertheless, Paul Slovic points out that "a device that succeeded in inducing insurance against the rarest hazards in our insurance experiments may also work for seat belts. If we can get people to change their perspective from that of a single trip [to] the risk of a serious accident over a longer time period— say a lifetime of driving—the accident probabilities . . . may be high enough to induce a general policy of always buckling up." Thus, while only about one in four thousand Americans will perish in a car crash each year, it is estimated that in the long run, one in sixty Americans born today will one day die on the road.

If seat-belt use remains abysmal, calls for the government to take safety decisions out of the hands of the public will undoubtedly increase. Appeals to reason will cease; reliance on technology will increase. Perhaps a new national consensus will grow on the efficacy of the air bag or automatic seat belt. Even conservative columnist George Will, no fan of regulation, has declared that he regrets "having once argued that government has no business requiring drivers to buy and use inexpensive devices that might save them from self-destruction. There is a pitiless abstractness and disrespect for life in such dogmatic respect for the right of consenting adults to behave in ways disastrous to themselves," he explains, adding that "too many children passengers are sacrificed on that altar," and that "a large part of the bill for the irrationality of individual drivers is paid by society.

"Most important," concludes Will, "society desensitizes itself by passively accepting so much carnage."

Close-Up (3)
EARTHQUAKES

"People in California know it's going to happen—but they don't think it will happen to them. The people in San Francisco think it will happen in Los Angeles, the people in Los Angeles think it will happen in San Francisco."
—Alex R. Cunningham, Director,
California Office of Emergency Services

When the big quake hits, the earth will shake for one, two, maybe three minutes. Buildings within a 250-mile-long by 100-mile-wide strip along the southern San Andreas Fault will rock, and tens of thousands will fall. Freeways in the Los Angeles metropolitan area and in Ventura, Santa Barbara, San Luis Obispo, and Kern counties will crack; overpasses and bridges will collapse. The smell of escaping natural gas will fill the air. Hundreds of fires will rage out of control and firefighters will be unable to reach them. Two of the three major aqueduct systems that cross the San Andreas Fault will be ruptured. At least one dam will break.

If the big quake comes during the middle of the night, perhaps only three thousand residents of southern California will die, according to a recent federal study. If it hits at midafternoon, at least twelve thousand will be dead and fifty thousand will be hospitalized. Property damage and repair costs may total $30 billion, with 200,000 people made homeless.

The victims will not wonder what hit them. They will know that it was the inevitable, great earthquake. Aware of this risk for years, they nevertheless chose to remain in Southern California.

In the late 1960s, when assorted seers, crackpots, and respected scientists began predicting that California was due for the "big quake" (as it is often referred to by expert and layman alike), most people, in and out of the state, reacted with a shrug or a smirk. Cultists had been making similar projections for years; disaster had not struck. It became a big joke. Maps went on sale showing Nevada as the new West Coast of the United States, with surfers frolicking on the beaches of Reno-on-the-Pacific. Californians quipped: "After the quake *we'll* stay put—the rest of the country will slide into the Atlantic."

The jokes diminished after the most destructive California quake of recent times (with a magnitude of 6.4 on the Richter scale) hit San Fernando, just north of Los Angeles, on February 9, 1971. The ground shook for just twelve seconds, but that was long enough to wreck twelve overpasses, a water filtration plant, and hundreds of houses. Damage exceeded half a billion dollars. Forty-seven people died when a Veterans Administration hospital collapsed. The Van Norman dam cracked, and if the ground had shaken just a little longer, it probably would have failed, driving one hundred thousand people from their homes in the valley below.

If the San Fernando quake was nature's way of warning the millions of people flocking to California, it failed. In the decade following the quake, the population of the state soared 19 percent, to over 24 million. (Closer to the epicenter, in the San Fernando Valley, the population rose 5 percent in the same period.)

Seismologists term the San Fernando quake a "major" event. The official expression for the quake that will soon strike Los Angeles (or another area of the state) is "great." It will be, perhaps, 900 times as strong as what struck San Fernando.

Every day California is rocked by several miniquakes, but only rarely do they cause damage. A recent pamphlet issued by the state Office of Emergency Services—the state agency responsible for earthquake preparedness—declared: "Earthquakes are as much a part of California as the Golden Gate Bridge, surfing or Hollywood." While one good shake is normally worth a thousand words of warning from experts, a federal report issued in January 1981 caused more of a stir in California than anything since San Fernando swayed.

The report was prompted by a trip President Carter took by helicopter over the Mount St. Helens disaster area in May 1980. The President was shocked when his top science adviser commented that the hundreds of square miles of wasteland below was a "scratch on the arm" compared to what a great quake could inflict on California. Carter asked the National Security Council to look into the matter.

A Federal Emergency Management Agency (FEMA) report, based on the National Security Council study, stated that "earth scientists unanimously agree" that the arrival of the "great" one was not a matter of if, but when. Scientists, the report said, had determined that at least eight great quakes had hit Southern California along the San Andreas Fault, thirty miles east of Los Angeles, in the past 1,200 years, with an average spacing of 140 years—plus or minus 30 years. Because the last great quake in Southern California occurred in 1857 (when the population of Los Angeles was 1,610), this means that the region became highly vulnerable again around 1967, and will continue to be held hostage by nature until the day it is widely laid waste (almost surely by 2027).

In other words, Los Angeles is due. It will remain so during the lifetime of most of the people now living in the area; and if the big quake does not hit soon, it will be, for the children of today's residents, overdue.

Geologists calculate that there is "well in excess" of a fifty-fifty chance that a "catastrophic" quake (an 8.3 on the Richter scale, equivalent to the 1906 quake which leveled half of San Francisco) will hit Southern California within the next thirty years, according to the FEMA report. There is, said FEMA, a 2 to 5 percent chance of this catastrophe occurring during any given year.

Five percent means that this year there may be a one-in-twenty chance that Californians (and to a lesser extent, the rest of the country) will have to cope with an event whose impact, the FEMA report said, "would surpass that of any natural disaster thus far experienced by the Nation. Indeed, the United States has not suffered any disaster of this magnitude on its own territory since the Civil War." Every year that passes uneventfully raises the odds that the great quake will take place the following year.

The Galveston flood of 1900 (which left six thousand dead) and Hurricane Agnes in 1972 ($3.5 billion in property damage) are the worst natural disasters to strike the United States to date. The highest death toll in an earthquake is seven hundred in San Francisco in 1906. (By comparison, the Tangshan earthquake of 1976 killed at least a quarter of a million Chinese.)

California has the misfortune of being located along the Pacific Basin, around which 80 percent of the world's major quakes—in Alaska in 1964 and Tokyo in 1923, to name just two—have taken place. Earthquakes happen when the immense "plates" that form the earth's surface shift along the natural boundaries where they meet. In California, the Pacific Plate, on which Los Angeles sits, meets the North American Plate, where San Francisco and most of the rest of North America reside, along the San Andreas and related fault systems. The plates move about as fast as fingernails grow—two inches every year. (In about ten million years San Francisco will be a suburb of Los Angeles.) They would like to move a little farther, but this movement is hindered by friction against the companion plate. This friction leads to strain, some of which is released through moderate quakes.

Geologists believe, however, that most of the strain can only be released through quakes that surpass 7.0 on the Richter scale. A quake occurs when the two plates break loose and suddenly slide twenty or thirty feet past each other. Great quakes release energy equivalent to 100 hundred-megaton bombs.

This often happens without warning. The much-ballyhooed science of predicting earthquakes is still in its infancy, despite some encouraging progress in countries (Japan, China, and the Soviet Union) that spend far more than the United States on this science. "We cannot report that we have found a complete solution to this difficult problem," Dr. Dallas Peck, director of the U.S. Geological Survey, told Congress in 1982. Scientists in the United States have fortunately not had a chance to observe a giant quake at close hand. This shortcoming will be remedied when the great California quake hits—but by then, of course, whatever lessons scientists learn from it will have come too late to help its victims.

The Los Angeles area is not the only section of California living in daily peril. The FEMA report sketched scenarios for several other expected quakes.

Northern California is twice cursed. One day soon a magnitude 8.3 quake along the Northern San Andreas Fault, or a magnitude 7.4 quake along the Hayward Fault, will rock the San Francisco Bay area. FEMA terms the likelihood of either occurrence in the next two to three decades as "moderate" (that is, a one in one hundred chance each year). These quakes would cause twice as much damage as the Southern San Andreas quake, but slightly fewer casualties, due to the lower population density of Northern California.

Taken as a whole, the FEMA report was pretty grim reading. Worse, perhaps, was FEMA's readiness review. Despite the predictions of the seers, the hints from San Fernando, and the warnings from geologists, "the Nation," FEMA said, "is essentially unprepared for the catastrophic earthquake," for "while current response plans and preparedness measures may be adequate for moderate earthquakes, Federal, State and local officials agree that preparations are woefully inadequate to cope with the damage and casualties from a catastrophic earthquake, and with the disruptions in communications, social fabric and government structure that may follow."

As an aide to a California legislator put it: "Earthquakes are not a political priority, unless they happen while you're in office." It has been said that government officials pray for two types of disasters: those so small that they go away on their own, and those so large that nobody can conceivably accuse the officials of unpreparedness or mismanagement.

More than a year after the release of the FEMA report, Senator Harrison Schmitt chaired a joint Senate–House subcommittee hearing on the subject. "Quite frankly, from what I have heard and what I know," said Schmitt at the conclusion of the hearing, "I do not think that we are prepared, and I just do not think we have in place the mechanism to respond." To get the government interested in this area, Schmitt observed, "it is going to take, potentially, the deaths of tens of thousands of Americans."

In May 1983 a magnitude 6.5 tremor rocked the tiny town of Coalinga, California, midway between Los Angeles and San Francisco. Coalinga's downtown was destroyed and forty-seven residents were injured. More troubling, however, was that the quake apparently indicated that the state was in even worse seismological shape than had been thought. The Coalinga quake was caused by a very large active fault, just off the San Andreas, which had not appeared on any California fault map. William L. Ellsworth of the U.S. Geological Survey in Menlo Park, California, said he was surprised to find such a fault "hidden in the basement," and admitted that "we don't know all the faults in California that can cause a magnitude 6 earthquake." Critics of California's nuclear power industry immediately raised questions about whether any nuclear plant in the state was safe.

This was no idle concern. In 1975, for example, a magnitude 5.7 quake struck Oroville, in a purportedly low-seismic portion of Northern California where a major dam had just been built. Fortunately, structural damage generally doesn't occur with quakes of less than 6.0 magnitude.

In 1983 the news grew ever grimmer for California residents. In a massive Earthquake Response Plan distributed by the Office of Emergency Services it was stated that a great San Andreas earthquake "is virtually certain within the next thirty years." And a draft report from a Seismic Safety Commission task force found that in regard to dealing with the big one to come, communications systems, training, planning, public awareness, and official leadership were all still quite inadequate to the task. (Officials in Los Angeles, however, were starting to enforce an ordinance requiring property owners to reinforce thousands of old masonry buildings downtown.)

Despite the inevitability of unpleasant surprises—and the almost equally inevitable failure of government at all levels to adequately prepare for an earthquake disaster—the citizens of California don't seem terribly alarmed. To use just one yardstick, it is estimated that less than five percent of homeowners in the state have taken out earthquake insurance (eight percent in the San Francisco and Los Angeles areas). To use another: reports of California residents moving out of the state because of quakes are rare indeed. (National studies show that natural hazards have little effect on where people decide to live.)

But if Californians are not taking out insurance, or taking flight, it is not, as is widely believed, because they are unaware of the dangers of a major quake. This may have been true twenty years ago, but after San Fernando, an increase in the number of earth tremors in the late 1970s, and then Coalinga in 1983—plus wide media coverage of various earthquake disaster scenarios—most Californians are well aware of what awaits them.

A 1981 report from the Institute for Social Science Research at the University of California, Los Angeles (UCLA), based on a survey of 1,450 residents of Southern California, noted that 43 percent thought there "probably or definitely" would be a significant quake in the region some time during the following year; very few felt there definitely would not be. This pessimistic sentiment, the report noted, went well beyond even the most chilling views expressed by experts. A vast majority of the sample wanted more earthquake coverage in the media. This, the report stated, contradicts the "widespread assumption" that most Californians deal with the big quake by "denial."

Why, then, do so few people in California seem truly alarmed by the prospect of a big quake? The UCLA report said it was necessary, in this regard, to distinguish between *awareness* and conspicuous concern. During the survey:

- Interviewers asked the 1,450 individuals who were surveyed to name "the three most important problems facing the residents of Southern California today." Even with three chances, only thirty-five mentioned quakes.
- The same 1,450 interviewees were asked if there was any problem endemic to Southern California that they would mention to a friend considering moving to the area. Only twenty-six mentioned earthquakes.
- Almost 20 percent of the sample said they believed that Southern California was a more hazardous place to live than other sections of the country—but nearly one-third felt it was less hazardous. And of the 287 people who felt it was more hazardous, only twenty-five mentioned quakes. Many of those who felt that it was less hazardous indicated that the earthquake threat was less severe than the tornadoes, hurricanes, winter storms, and floods that confront people in many other parts of the country.
- Twenty-seven percent of the sample said they were "very frightened" about the possibility of experiencing a dam-

aging quake; only 14 percent were not frightened at all. But, when asked what they would do if they were "certain" that a damaging quake was about to occur at a specific time near where they lived or worked, a substantial 34 percent said they would "go on as usual" with their life; only 29 percent said they would try to get as far away as possible.

- Few indicated that they had followed any instructions from officials on how to prepare for a major quake (such as by storing food and water). Those who expressed a fear of earthquakes were somewhat better prepared than average, "but the most fearful fifth of the population," said the report, "have done less"—which, it continued, "supports a common psychological principle that moderate fear is productive but that very high levels of fear can be counterproductive."

These results indicated to the UCLA researchers that even after learning much about various earthquake warnings, "very few people living in earthquake country are preoccupied with the threat to their safety." Problems such as crime, the cost of living, taxes, unemployment, smog and pollution, transportation, overcrowding, and education, said the report, all came to people's minds before any thought of earthquake danger. Perhaps, in one respect, this is a good thing. For years, officials in California were concerned that so few residents were taking out earthquake insurance. Now some are worried that *too many* will do so. Insurance companies have already extended almost $10 billion worth of coverage in San Francisco and Los Angeles and fear that they will be bankrupted by the Great One.

There is also the "misery-loves-company" factor. A study done by Jennifer McKay of the University of Melbourne revealed that when shown a map dominated by a close-up view of their region, the residents of an Australian flood plain generally had a "We're all in this together, so why worry?" response. When shown a map in which their flood plain appeared as a small zone within a large, hazard-free area, however, they felt extremely alienated ("Why *me* out of all these people?").

Residents with this feeling, the study found, were more likely to undertake some form of preparation.

Given this, one imagines that the residents of Southern California would be much more worried about the coming quake if their counterparts in the north were not also imperiled.

During a five-day period in January 1983, the resort area around Mammoth Lakes, California, was rocked by a "swarm" of three thousand quakes. Several measured as strong as 5.6 on the Richter scale, knocking out power and causing minor damage. Weekend ski business at the resort was reported down only slightly.

Alex R. Cunningham, director of the California Office of Emergency Services, feels that the reason most Californians ignore the earthquake threat is that "nobody likes to think about disaster." Cunningham, forty-six, does a lot of public speaking around the state. Frequently people respond to his remarks by saying, "Well, I could get hit by a car tomorrow." A transplanted East Coaster who recalls his own worry about earthquakes when he moved to California in 1959 ("I thought the sidewalks would open up and swallow me"), Cunningham believes that most people "feel there's nothing they can do about it," and that somehow, "Big Brother is going to come in and make everything right." They don't realize, Cunningham points out, that there will be general chaos, and that they have to prepare to take care of themselves for at least forty-eight hours.

Cunningham complains that his job of putting quake consciousness on the "front burner" is made tougher by the lack of certainty that surrounds this issue. He explains that if he went to the governor of California or the mayor of Los Angeles and told them that there was certain to be a big earthquake in, say, two years, "Businesses wouldn't locate here, and those that were in the area would relocate. And then what if it didn't happen?"

Even if seismologists could reliably give two or three days' warning of a quake holocaust, says Cunningham, it wouldn't save the city because in that time, "you still couldn't reinforce the buildings." Nevertheless, he sees some hope. Someday a prediction will be made, "people will pooh-pooh it, the earthquake will happen, and then they'd listen the next time."

Unlike Cunningham, who lives and works in relatively safe Sacramento, Gina Lobaco, a writer, her husband Bill Blum, an attorney, and their infant son, Max, live in a quake-prone section of the state. Although Bill is a newcomer to California, Gina has lived in the Los Angeles area almost continuously since 1957. While schoolchildren in other states were ducking under their desks in civil defense drills, she and her classmates were practicing the same maneuver in preparation for earthquakes.

When the San Fernando quake hit in 1971, Gina was living about twenty-five miles from the epicenter. The quake struck at 6 A.M. She recalls that her father woke up, lit a cigarette, and said to her mother: "Well, I guess this is it—the end of the world." But her house suffered no damage, and everyone she knew, although shaken up by the experience, "went to work that day."

Now, whenever an earth tremor occurs—which is three or four times a year—Gina calls her husband or he calls her to ask whether the other felt the rumbling. Once, when a tremor hit while Bill was talking to Gina on the telephone from an office in a tall building, he kept slipping out of his seat. Sometimes she awakes at night and "panics" when she feels the bed moving, thinking it's a quake. So far, it has always turned out to be Bill, simply turning over in his sleep.

A map printed in a local newspaper recently told them that the Eagle Rock section of Los Angeles (near Pasadena) where they live is built over a fault line connected to the San Andreas. Bill and Gina have had earthquake insurance—for which they pay $200 annually—for some time, but this is the only precautionary step they have taken. They ignore repeated instructions (on radio and television, and even on a separate page in their local telephone directory) to prepare for the worst.

"I think about earthquakes a lot," says Gina, "especially after I've heard something about them on the news. And whenever a slippage in the San Andreas is reported, I think, 'Well, it's time to move East—you have to shovel snow but at least there's no earthquakes.' Then I just forget about it. I'm aware that we're due for a big one, but I guess I feel my house won't fall down around me. I feel it has survived so many tremors that it would stand up in the big one. That doesn't follow, of course. I worry more about it when I'm in large buildings but I feel they are seismically sound, so it's okay.

"I guess I just don't have a siege, or a survivalist, mentality," she explains, "maybe because I've never experienced a major disaster. I have no notion what it would be like. I've read all the disaster scenarios but I block them out, I guess. Every so often I think we should get a first-aid kit together, or something, but then we don't. I mean, we're just now getting smoke detectors."

Some people in Southern California actually find earthquakes kind of exciting, Gina reports. They tell her it is "wonderful" or "amazing" how there's this huge movement in the earth— "this static mass becoming animated, taking on life." Some people, Gina says, pull "dirty tricks" on their spouses by jiggling their toes under the covers at night, trying to make them think that an earthquake is arriving; she's done it herself.

One time, when Bill and Gina were visiting Bill's parents in Pennsylvania, a hurricane came by during the night. Bill slept right through it but Gina was terrified. And so, like other confirmed Californians, Gina will continue to take her chances in Los Angeles. "I don't know anyone who's moved out," Gina says. "It's so nice here. California is a real nice place to live."

Most Californians who have made their peace with the possibility of being physically harmed by the great quake are probably unaware, however, of the devastating long-range aftereffects of this event.

When the San Andreas Fault shifts, it will be almost as though several nuclear bombs had fallen on Southern California (without the deadly radiation). The area's electrical power supply will be cut in half. Telephone service will be severely disrupted. Many hospitals will be destroyed and those still standing will be brutally overtaxed. There will be no place to put the hundreds of thousands of homeless. Older, unreinforced brick buildings (an estimated eight thousand in Los Angeles) can be expected to be "poorly resistant to earthquakes," according to FEMA; this would include buildings put up as recently as the late 1960s. Even some newer, "quakeproof" buildings may not survive. (An eight-year-old, $1.8 million county building collapsed during the moderate Imperial Valley quake of 1979.) If aqueducts rupture it will take three to six months to restore water supplies. The impact on transportation, the FEMA report said, could be "massive," with railroads (and pipelines) crossing the quake zone in an east-west direction suffering "initial losses . . . approaching 100 percent"; highways, airports, and harbors would suffer 15 to 30 percent damage.

But that's not all: The economy of the wealthiest portion of the nation's richest state will be crippled indefinitely. Many of the countless factories and warehouses that fall will never rise again. Private insurance companies will be unable to pay the unprecedented damages that will occur, and will eventually default; the federal government will be hard pressed to make up the billions of dollars outstanding. Thousands of the unemployed will wander the streets. Bank officials warned recently that a quake could totally knock out their computer systems, causing severe financial ramifications across the state, and country, within days.

Because 10 percent of the industrial resources of the United States are based in California—and 85 percent of the state's resources are located in the twenty-one counties facing a major quake disaster—any enormous rumble will also be a national disaster. Much of the nation's electronics industry, for example, is located in those counties, which produce 40 percent of America's semiconductors, 25 percent of its computers, and 21 percent of its optical instruments. The economic aftershocks of the California quake could equal those that reverberated from the Dust Bowl in the 1930s—which, ironically, caused so many people to migrate to this earthquake-cursed Promised Land to begin with.

Gina Lobaco admits that she has "never thought about" the chaotic aftermath of a quake. "It's never really hit me," she says, "that earthquakes could affect the whole social and economic fabric of the region. I hear a lot more discussion among people about the fantasy aspect—you know, whether it's physically possible for the state to slide right off into the ocean. You think more about that stuff than the economic fallout."

People living in other parts of the United States, however, shouldn't feel smug about California's earthquake problem, since many are themselves not immune to such a disaster. Portions of thirty-nine states are at risk for major quakes; within these areas live seventy million people.

Particularly imperiled are New England (including the Boston area), Salt Lake City, Charleston, South Carolina (where a major quake caused vast devastation in 1886), and a seven-state section of the Midwest. During the next two hundred fifty years, according to U.S. Geological Survey director Dallas Peck, "every part of the United States except for a small portion of southern Texas, Louisiana, Mississippi, Alabama, and Florida could experience levels of . . . shaking that could damage some structures that are not designed and constructed to be earthquake resistant. The building wealth exposed to the ground-shaking hazard is in the trillions of dollars."

In fact, 1983 was the Year of the Quake across America. There was the Coalinga tremor, of course, and a 7.3 magnitude quake that shook Idaho, killing two children—the strongest quake since 1952 in the continental United States and the first earthquake fatalities in the continental United States since 1971. A 5.2 shocker hit northern New York, setting off tremors throughout the Northeast. The U.S. Geological Survey reported 342 quakes that were "felt" during the year, and as in previous years California led all others with 119 "felt."

Even though the likelihood of a big quake on the West Coast is far greater than it is in the East, when a big one hits outside California the damage will probably be much greater. The earth's crust east of the Rocky Mountains propagates seismic waves much better than that west of the Rockies. Therefore, waves from a magnitude 7.0 quake in the East, for example, will be felt over a much wider area than from a similar quake in the West. Another regional difference is that while the major fault lines in the West are visible from the air, in the East they are buried under several kilometers of sediment and cannot readily be mapped by geologists.

Prompted by new seismologic studies, the Nuclear Regulatory Commission recently began reassessing the earthquake threat to every nuclear power plant east of the Appalachians. The design standards for nuclear plants in the East had assumed that a magnitude 7.0 Charleston-type quake could happen only in that region of South Carolina. Now, according to James Devine, a Geological Survey official, "it has become evident that the general geologic structure of the Charleston region can be found at other locales within the Eastern Seaboard."

Aside from California, the area most likely to suffer a calamitous quake, the Geological Survey announced in 1983, comprises parts of Indiana, Illinois, Missouri, Kentucky, Tennessee, Arkansas, and Mississippi. In fact, this country's worst quakes, termed "unique in their awesomeness" by one expert, occurred in 1811 and 1812 near New Madrid, Missouri. A series of three quakes, which according to modern estimates would have hit 8.0 and possibly 10.0 on the Richter scale (if the scale had existed then), shook a million-square-mile section of the country. Shocks were felt in Boston, 1,100 miles from the epicenter, and made church bells ring in Washington, D.C. Trees split in two; the Mississippi and Ohio rivers flowed backward; the ground sank in northwestern Tennessee to form Reelfoot Lake, five miles wide and eighteen miles long. Damaging aftershocks were felt for years. This sparked the first serious study of earthquakes in the United States.

Only one person died in the sparsely populated Midwest during the 1811–1812 quakes, but a recurrence would kill thousands. Over twenty-three thousand people live in the city of New Madrid today, and the cities of Little Rock, Memphis, Evansville, and St. Louis, among others, all lie close to the New Madrid Fault. An earlier U.S. Geological Survey study pointed out that a repeat-quake would cause losses in the range of twenty thousand dead and $45 billion damage.

This may significantly understate the outcome. A FEMA study in 1982 noted that since few dams in this region have been designed to withstand earthquakes, "a substantial number of dam failures can reasonably be expected in any major earthquake which affects the Central United States." The FEMA report listed over a thousand dams in Missouri alone which were either "unsafe" or considered a "high hazard" because their failure would cause substantial loss of life.

"The Central United States has little history of legislation at the State level favorable to an improvement in the seismic safety requirements in building codes, professional practices or land use regulation," FEMA officials Charles C. Thiel, Jr., and Ugo Morelli have observed. In April 1984 FEMA provided a $300,000 grant to the seven imperiled states to develop plans for dealing with the major earthquake scientists now said was likely to occur within twenty-five years.

Yet there is no evidence that residents in this area are at all alarmed about the possibility of a catastrophic disaster. In his book *The New Madrid Earthquakes*, published in 1981, James L. Penick noted that although tremors are regularly recorded by instruments in the region, many of the twelve million people now inhabiting the area damaged in the 1811–1812 quakes feel a false sense of security. Perhaps they feel they have enough (frequent floods and tornadoes) to worry about already. And, as Thiel and Morelli declared: " 'Everyone' in California knows that they are in 'earthquake country'; however, few in the Central United States realize that major damaging earthquakes can occur in their area and that steps to reduce vulnerability can be taken."

2
The
"Informed"
Public

We like to think that we're well informed about most factors that affect us greatly, but are we really? Our judgments and decisions about risks are largely based, after all, on information given to us in a variety of ways and from diverse sources that often color it to suit their vested interests.

What, for example, was the public to make of the risks associated with dioxin in the wake of the flurry of controversy over this substance in the summer of 1983? The brouhaha began when the American Medical Association's house of delegates approved a resolution accusing the news media of making dioxin the subject of a "witch hunt" through "hysterical malreporting" that "ignorantly damaged" the lives of residents of contaminated areas (such as Times Beach, Missouri). The AMA voted to conduct a publicity campaign to alleviate public concern about the chemical, which had become the victim of "rumors, hearsay and unconfirmed, unscientific reports." A few days later President Reagan offered the opinion that reports about dioxin had "frightened a good number of people unnecessarily." Two respected publications, Science and Chemical & Engineering News, suggested that dioxin was really not much of a hazard.

This sparked an angry response. Noted cancer expert (with a bias of his own) Dr. Samuel S. Epstein, for example, called the AMA's position a "travesty" fully consistent with the association's "past record of ignorance and poorly developed expertise in these areas." The uproar forced the AMA's leadership to backpedal. One representative of the group told a Congressional subcommittee that "the AMA does not pooh-pooh dioxin" and that dioxin was indeed a potentially serious health problem. A few days later, three top Reagan Administration public health officials told Congress that new evidence linked dioxin to cancer.

But what impression of dioxin's risk had been left on the public after the controversy died down? After all, the President of the United States and the nation's leading medical association had instructed citizens to allay their fears—an effect that even the AMA's partial recantation did not wipe out.

Fortunately, dioxin is not a consumer product available on supermarket shelves; most people will never have to decide whether or not to expose themselves to it. But what about the many cases in which the public does have to make such a choice in the face of conflicting testimony or unresolved debate?

The manufacturers of consumer products, foods, and drugs usually generate advertising that stresses the appealing characteristics of their products while leaving the risks unmentioned. At the same time, scientists who seek to warn the public about such risks often fail because their messages are worded in technical language that few can understand. So manufacturers are clearly at an advantage.

A case involving the artificial sweetener saccharin demonstrates how technical factors and the debates that surround them confuse the public into passively accepting (and sometimes even embracing) a potential risk. In 1978 the Institute of Medicine and National Research Council/National Academy of Sciences jointly published the following outline in their report entitled "Saccharin: Technical Assessment of Risks and Benefits":

1. In rats saccharin is a carcinogen of *low* potency relative to other carcinogens.
2. But in addition to acting by itself, saccharin promotes the cancer-causing effects of some other carcinogenic compounds in rats.
3. Whether as initiator or promoter, saccharin must be

viewed as a potential carcinogen in humans, but one of low potency in comparison to other carcinogens. Although saccharin would be expected to be of low potency in humans, even low risks applied to a large number of exposed persons may lead to public health concerns.

4. The state of the art in extrapolation does not permit *confident* estimation of the potency of saccharin as a cause of cancer in humans.

5. Essentially, there is no scientific support for the health benefits of saccharin.

6. Although alleged psychological benefits of the use of saccharin and other non-nutritive sweeteners cannot be evaluated at this time, it is evident that segments of the population regard the substance as desirable.

7. The observation that young children are becoming increasingly greater consumers of saccharin suggests that public health officials should take a prudent course of action since there has been insufficient time for the possible effects of this greater consumption to be manifest.

For ordinary people to reach a decision on whether or not they should use saccharin requires understanding such terms as "initiator," "promoter," "carcinogen," "potency," "extrapolation," and "risk level." These must then be weighed against the benefits and appeal of saccharin, both physical and psychological. It is small wonder that the public is often in a quandary.

Despite gaps in information and the dispute over the data on saccharin, an amazing number of Americans rushed to defend the sweetener when, in 1977, the Food and Drug Administration proposed banning it. The FDA's action was based on the Delaney Clause of the Federal Food, Drug and Cosmetic Act, which mandates the removal of any food additive shown to be carcinogenic in animals or humans. The Delaney Clause, named after Congressman James J. Delaney of New York, was enacted in 1958 and leaves no room for doubt: If *any* food additive, ingested in *any* amount by *any* animal, no matter for how long, brings about *any* form of cancer, it must be banned. The clause itself has been a matter of heated controversy for many years, and even the FDA, which derives considerable power from it, does not always find that it works in the public interest.

One absurdity created by the Delaney Clause is that it is applied only to *new* food additives and not to those already in use at the time it was enacted. This latter group comprises a portion of the FDA's GRAS ("Generally Recognized As Safe") list of food additives that had been previously approved. The fact that the law does not apply to these substances means, in effect, that Congress has sanctioned the use of established cancer-causing additives, such as Red Dye #3, but not newer ones. In a report to the FDA in 1980, a group of independent scientists who had reviewed 415 substances on the GRAS list could only declare 305 of these substances safe. They called for restrictions on the use of five of the substances, found "uncertainties" about the safety of nineteen others, and suggested further study of the remaining ones.

When the FDA published its plans to ban saccharin, congressmen were deluged with mail from people decrying this move. Some cardiologists, bariatric physicians ("fat doctors"), and diabetes specialists publicly predicted all sorts of dire consequences for their patients if saccharin were banned from soft drinks (its most popular use). A leading New York cardiologist warned that the replacement of ten billion cans of diet soda with regular soda would lead to an extra two trillion calories being consumed each year—a calculation which wrongly assumed that people would continue to drink just as much soda— which, he said, could lead to twenty-five thousand extra heart attacks annually. Although informed medical scientists found it difficult to accept such conclusions, they made an enormous impression on the general public.

Entering the saccharin fray, a spokesman for the American Diabetes Association predicted more serious complications in diabetic children if they didn't have artificially sweetened gum and soft drinks; among these he listed blindness, strokes, heart attacks, and kidney disease. What the ADA spokesman didn't say was that these are the usual complications of diabetes, and can occur whether or not one chews artificially sweetened gum or drinks low-calorie soda. He also failed to note that in Canada, where saccharin-sweetened soda had been banned, no impact had been observed on the health of diabetics. Nor were matters clarified by a former director of the National Cancer Institute, who suggested that banning saccharin might *cause* cancer by increasing obesity, claimed by some to be a cancer risk factor.

Dr. Sidney M. Wolfe of the Health Research Group (founded by Ralph Nader) characterized these statements as irresponsible. Wolfe argued that they were predicated on two basically false assumptions: (1) that saccharin helps people lose weight or prevents them from gaining it (for neither of which any evidence exists); and (2) that the $4-billion-a-year diet soda industry would be unable to come up with a low- or no-calorie substitute for saccharin or cyclamates (which had been banned in 1969, another victim of the Delaney Clause). Wolfe convincingly argued that the second assumption was a self-fulfilling prophecy, because it stirred thousands of sincere consumers and doctors into protesting the ban. In fact, within four years, a new artificial sweetener called Equal, containing the substance known as aspartame, had surpassed Sweet 'n Low as the leader in the "tabletop" diet sweetener field, and was being widely used in soft drinks. (However, there are now some questions about the safety of aspartame.)

What Wolfe didn't say is that some of saccharin's scientific and medical defenders were the paid allies of the Calorie Control Council, a trade association of some sixty American soft-drink companies. Others were physicians who simply stated their own opinions based on scanty knowledge—although the impact of their faulty judgments was enormous and pleased the Calorie Control Council and its members no end.

Bowing to pressure generated by many sincere but ill-informed citizens, Congress passed the Saccharin and Labeling Act of 1977, which effectively put a moratorium on the proposed ban, pending further study. In August 1981 President Reagan, sympathetic to manufacturers' needs and favoring less government regulation and control of industry, extended the moratorium.

The government decision to ban saccharin has since evolved into a controversy involving several issues. Among these are risk estimation, contradictory epidemiological and experimental evidence, and individual freedom versus government control. Whether real or contrived, such controversies typically confuse the public completely. The failure to resolve them leads to the continued marketing of risky products. "If you were a soft-drink manufacturer and had the freedom of choice between two ingredients," Wolfe asked, "one which cost about six cents per gallon of syrup concentrate [saccharin] and the other about a dollar per gallon [sugar], but the syrup would sell for the same price with either ingredient, which would you choose?"

Once the aura of controversy surrounds an issue, it becomes increasingly difficult to find out where the truth lies. Each side musters ever more evidence and favorable scientific opinion. It then becomes a matter of which expert you believe. Making a decision isn't easy because experts are quite unequal in many ways, including their ethical standards and professional ability—characteristics the public cannot readily discern. Moreover, the experts' continuous opining on issues not easily defined in the first place serves to make the controversy interminable. Often the end result, as a recent New York Times headline pictured it, reads: "1 Expert + 1 Expert = 0."

While we try to keep an open mind on controversial topics such as the safety of saccharin, surprisingly often we embrace the opinions of a single expert—at least until he or she is proved wrong.

A few years ago, for example, Dr. Ben F. Feingold, a California pediatric allergist and retired staff member of the Kaiser-Permanente Medical Center in San Francisco, made the claim that at least 40 percent of children with hyperkinesis (hyper activity) and consequent learning disabilities showed a marked improvement when artificial food colors and flavors, as well as naturally occurring aspirin compounds, were removed from their diet. In what was to become the extremely popular book entitled Why Your Child Is Hyperactive, Dr. Feingold published a long list of fruits and vegetables that should be excluded from the diet because they contained what he termed "natural salicylates." He also recommended the avoidance of bakery products, chewing gums, some ice creams and soft drinks, and some gelatin products because they contain artificial food colors.

Dr. Feingold's book, and the Feingold Diet, found a ready

audience among those individuals beset by the problem of hyperactive children. One simply had to avoid feeding them certain foods and the symptoms would go away!

To some physicians, teachers, parents, and psychologists, it seemed too good to be true, and they maintained a healthy level of skepticism. But others embraced it completely, including some physicians. They did so not so much because they were adequately informed, but because Dr. Feingold's diet was an appealing solution to a painfully intractable problem. Essentially, *he told people what they wanted to hear*. In so doing, however, he actually misinformed them (as subsequent evidence showed)—not intentionally, but rather because his observations were based on faulty study design.

Between 1975 and 1980, dozens of scientific investigators conducted well-designed human experiments in an attempt to corroborate Dr. Feingold's findings. These studies, which collectively cost close to a million dollars, were funded by diverse groups, including the Nutrition Foundation, the FDA, and the National Institute of Education. In these experiments children were divided into two groups: those fed foods containing the substances Dr. Feingold claimed produced hyperkinesis and those fed "placebo foods."

In October 1980, the National Advisory Committee on Hyperkinesis and Food Additives, which had coordinated the research, announced the results of these experiments: No regular and consistent pattern of improvement was found in hyperactive children fed the Feingold Diet. And researchers were unable to detect further deterioration in children who were fed foods that Dr. Feingold had claimed caused hyperkinesis.

Of great interest, however, was the finding that some children in the group fed the Feingold Diet *did* temporarily improve. This common phenomenon is known as the "placebo effect," whereby people given a fake pill, for example, may improve to the same extent as those given a real medication. But the placebo effect is neither consistent nor sustained.

Because Dr. Feingold did not use controls in his studies, the "placebo effect" in large measure explains his findings. He merely observed what happened to hyperkinetic children when they were fed his elimination diet, not what happened to an equal group fed a regular diet disguised as his own—a sincere and honest mistake, yet one that misinformed and misled many individuals and cost a million dollars to clarify.

The National Committee on Hyperkinesis and Food Additives didn't suggest a ban on the Feingold Diet, since it appears to be harmless and has a slight placebo effect on some children. But the committee cautioned that it should only be used in conjunction with other forms of treatment. This conclusion left a wedge in the door, so to speak, and many have used it to sustain their belief in the scientific soundness of the diet. This is understandable because the parents of hyperactive children have few other treatments on which to pin their hopes. And, they reason, the experts may be wrong after all, so why not try it? In so doing, these parents make an individual choice based on pressing personal need, not factual information.

Perhaps even more disillusioning for the public than wrong advice is expert opinion that seems to change with the seasons. "No assertion has a shelf life of more than eleven months," *Newsweek* columnist Meg Greenfield complained in the wake of reports that fatty fish and shellfish, once banned from low-cholesterol diets, were now thought to actually reduce cholesterol levels in the blood. The public now reacts to fresh warnings, Greenfield suggested, by ceasing to pay attention to them. "The fact is," she wrote, "that the public, far from being gullible and vulnerable and the rest, is damned crafty and exploitative in these matters—itself expert at manipulation. By now it suspects that what is banned today is likely to be administered intravenously tomorrow.

"It has learned," Greenfield wrote, "to invert the findings where necessary, to discount the permanence of any of them, to make the unwholesome, illogical best of whatever it is told. Nixon is back. Everything is back. Twinkies will be back. They will be telling us to put more salt on them." A short while later it was reported that the artificial sweetener cyclamate was undergoing an FDA rereview and may "be back" in the near future.

In many cases, however, even the most accurate information, presented in the most consistent fashion, does not succeed in sparking preventive action and reducing risk. The public often rejects good advice, or chooses to ignore it entirely. Therefore, the thesis of "just-give-the-public-the-facts-and-they-will-act-wisely" is not the answer to the gravest risk-reduction questions.

Perhaps the most obvious evidence of this is the rather un-derwhelming effect of the warning label printed on every pack of cigarettes purchased in America. But this is by no means an aberrant example. The federal government spends millions of dollars annually on public service and consumer education announcements, often to little avail, and consumers routinely make little use of the safety and health information that comes with products.

A study at Johns Hopkins University in which a group of more than one hundred women were put through an intensive course on home safety supports this. Investigators who later visited the women's homes found no difference in safety prac-tices between them and a control group.

Such reports have sparked calls, from both the political left and right, to reduce federal expenditures on consumer educa-tion. Conservatives feel it is wasted money that could be saved; liberals want the money channeled into programs that mandate (rather than suggest) consumer action, such as installing air bags in cars rather than pleading with people to wear seat belts. "Safety campaigns may work to get people to perform a one-time act, like buying a smoke detector," says R. David Pittle of the Consumers Union, "but it is far more cost-effective to change a product than to change the long-term behavior of millions of consumers."

Why do people shun good advice? Do we receive so much bad advice over the years that we become practically immune to well-meaning suggestions? As in many areas of risk analysis, it is revealing in this instance to examine how people behave when it comes to buying insurance.

Many recent studies—and the popularity of insurance plans with small "deductibles"—indicate that people are more likely to insure themselves or their property against high-probability/low-loss hazards than against low-probability/high-loss haz-ards, as we noted earlier. These results run counter to the theory that insurance should be bought to protect against losses too great to bear, rather than for small losses that can be paid out of pocket. Yet people seem to prefer protecting themselves against minor hazards that are relatively likely to happen. This seems to reflect the view that a person can worry about (and protect oneself against) only a limited number of things; if we didn't ignore low-probability threats, we would spend our entire lives in a state of alarm or paranoia.

But insurance studies also indicate that we often make misguided choices in deciding *which* events to guard against. There are many reasons for this. One common mistake is believing that lightning (or some other hazard) can't strike twice in the same spot—or at least not right away. This, of course, is a total fallacy: the odds of any random occurrence happening twice in short order are slim, but after an event has happened once it is usually just as likely to happen again soon. Even residents of flood plains—who know (or should know) that storms can hit and the river overflow at any time—seem to feel after a major flood that they won't have to worry about another one happening again for a long, long time. They are firm believers in a "law of averages" that is, in the short run, quite mythical.

Another reason for faulty behavior is that people derive comfort from a low level of danger directly related to some single action: one cigarette, for example, never gave anyone lung cancer. The percentage of people who stopped smoking would increase dramatically, however, if smokers stopped thinking only of their next cigarette and began considering the risks inherent in a lifetime of lighting up. Chronic drug abusers, after safely shooting up or snorting some substance several dozen times, begin to feel like Superman—until that fatal moment when they take one hit too many.

A third explanation for rational but foolhardy behavior concerns something known as "availability biases," which distort people's perceptions. This simply means that when we make a judgment about the frequency or probability of some event, we often base that judgment either on recollections of past instances of the event or on how easily we can *imagine* the event happening. Using these "cues" is usually good empirical strategy; we *can* recall frequent harm more easily than infrequent harm, and we can imagine some likely danger much more clearly than a long-shot threat. But these cues are also affected by factors unrelated to actual likelihood: thus, for example, our remembrance of more recent, more vivid, or more emotionally affecting experiences tends to be out of proportion to their actual significance or frequency. This can lead to the severe over- or underestimation of certain risks.

We are also influenced by highly charged reports in the media, which tend to affect our imagination vividly (and therefore lead us to conclude that some events are more likely to happen than they really are). Many would-be visitors to our national parks, for instance, are scared off by their fear of being attacked by grizzly bears, the subject of some horrific stories in the press. In fact, only a handful of park visitors have been killed by bears, and according to some calculations would-be visitors are more likely to die in an automobile accident en route to the park than from an encounter with a grizzly.

All of this means that while we all make some very conscious decisions about which risks to avoid and which to accept, we are also very much at the mercy of what we are led to believe. As Paul Slovic, a leading researcher in the risk field, has put it: "The fact that subtle differences in how risks are presented can have big effects on how they are perceived suggests that people who present risks to the public have considerable ability to manipulate perceptions."

In the final analysis, people are free to believe what they wish, but frequently they tend to believe what suits their needs or appeals to their prejudices. Thus, no matter how valid or objective some piece of information is, most persons tend to modify it, and some even distort it. This leads to faulty decisions and judgments. Thus, rather than being based on the exercise of free will and intellect in the face of valid and objective evidence, the apparent willful acceptance of risk, or its emotional rejection, can be the product of misguided judgment applied to misinformation.

Close-Up
SATURATED FATS AND CHOLESTEROL

"If you're waiting to change your diet until there's an airtight case against fats and cholesterol, you may die waiting. . . . Sooner or later, you will have to make your decision based on less than the best possible evidence. . . ."

—Jane Brody,
Jane Brody's Nutrition Book

Why aren't more people aware of the facts about cholesterol, saturated fats, and heart disease? Vast portions of the public are either uninformed, misinformed, or (more frequently) confused about what they perceive as controversy over these facts. The sources of this confusion have been the chief purveyors of saturated fat and cholesterol in our diet—the dairy, egg, and meat industries and their hired scientific allies.

Each year approximately one million Americans suffer heart attacks, and 650,000 die as a result. Of those who die, one hundred fifty thousand are under sixty-five years of age. These figures reveal nothing of the millions of people who live with angina and other complications of coronary heart disease. What the figures do tell us is that coronary heart disease is epidemic in the United States.

Coronary heart disease usually results from severe atherosclerosis (hardening) of the coronary arteries, which narrows and eventually closes off these vital blood vessels. And the overwhelming evidence is that along with risk factors such as smoking and hypertension, diets high in saturated fats and a constituent of fats known as cholesterol lead directly to atherosclerosis and coronary heart disease.

Clinical and pathologic studies and animal experiments unequivocally show that saturated fat and cholesterol in the diet— found in large quantities in dairy products, meat, and eggs— increase the levels of cholesterol in our blood. The saturated fats we eat are converted by our bodies' chemical processes into cholesterol. By contrast, polyunsaturated fats, found in a number of vegetable oils and margarines, actually lower our blood cholesterol levels.

High cholesterol levels play a key role in the formation of "plaques" on the linings of our arteries. These plaques are yellowish patches consisting mainly of fats and cholesterol from the blood. The higher the blood cholesterol level, the greater the rate of plaque formation. As the plaques grow they increasingly block the flow of blood through the arteries, leading to strokes and heart attacks. Atherosclerosis and arteriosclerosis are the two terms most often used to describe the condition that results when arteries become heavily blocked and damaged by plaques.

The good news is that in many people, atherosclerosis and coronary heart disease can be prevented and controlled. Time and again it has been shown that those who limit their intake of cholesterol and saturated fats greatly reduce their risk of developing these problems. As Americans have cut back their consumption of eggs, milk, and animal fats (as well as cigarettes) in the past two decades, there has been a 4 to 5 percent decrease in the average blood cholesterol level of adults. Since 1968 deaths from cardiovascular disease have declined steadily.

But Americans still consume roughly twice as much cholesterol as they should—450 milligrams each day—and several hundred thousand continue to die every year from cardiovascular ailments. By some estimates the average American consumes the equivalent of a stick of butter a day in fat and cholesterol.

The evidence linking atherosclerosis to diets rich in saturated fat and cholesterol has been gathering since the early part of the century. In 1908 two Russian scientists discovered that rabbits fed milk fat and eggs developed atherosclerosis. Nikolai Anitschkow, another Russian scientist who followed up on this work, demonstrated that cholesterol caused atherosclerosis in rabbits.

But in 1913, when malnutrition and vitamin deficiencies were rampant in the United States, two University of Wisconsin researchers, Dr. Elmer McCollum and Marguerite David, discovered vitamin A, the first known vitamin, in egg yolks and milk fat. McCollum and David charted a course for American nutrition that saw most foods (except sugars) as containing nothing but beneficial nutrients.

So it is not surprising that the findings of the Russian scientists were ignored. While relatively few people in that era lived long enough to develop atherosclerosis, heart attacks, and strokes, hundreds of thousands of younger Americans were suffering from vitamin-deficiency diseases such as rickets. American nutritionists spent their time, therefore, finding new vitamins and publishing dietary advice.

In the short run, this saved hundreds of thousands of lives. However it also had long-range effects as well. Meat, eggs, and milk products became the darlings of the nutritionists because they contained so many nutrients and vitamins. Nutrition science and research became inextricably intertwined with farmers who produced food. Farm groups financially supported research that not only led to the establishment of human nutritional needs, but also an improvement in livestock quality to meet those needs. This helped farmers develop everything from better-egg-laying hens to cows that produced higher volumes of milk. The spinoff for the farmers was that they were able to more readily sell products now hailed by nutritionists as essential for human growth and development. Naturally, farmers and nutritionists became very good friends.

Eventually the government arrived on the scene, eager to provide the public with nutrition advice derived from research largely supported by farm groups. The U.S. Department of Agriculture eventually developed a list of foods known as the Basic Seven. Each food on the list contained one or several essential nutrients. At the time, everyone seemed happy with this approach—nutritionists, mothers and children, the government, and the farmers and food processors.

By 1956 new knowledge had reduced the Basic Seven to a Basic Four—meats, fruits and vegetables, dairy products, and cereals. At this time, the National Dairy Council, with the government's permission, became the largest and most important provider of nutrition education in the country. Its staff, situated in almost 125 cities in the United States, produces curricular materials for schools, nutrition columns for newspapers, and television and radio announcements. That the Dairy Council can still convincingly promote saturated fat and cholesterol-rich diets reflects both the credibility it built in the days before the link between these elements and atherosclerosis was known, and public uncertainty concerning this subject today.

The government has long sustained the interests of the egg and meat industries. One small example of this is the system of meat grading used by the U.S. Department of Agriculture. Meats are graded according to their fat content—with fattier meats given *higher* grades. (A high-grade meat may not always look fatty to the naked eye, but it is.)

This happy relationship among nutritionists, farmers, and the government, born in the early part of this century, grew as the years passed. Nutrition became a quiet field to work in, devoid of excitement and controversy, and leadership positions in it frequently fell to nonphysician scientists who were amply supported by research grants from the meat, dairy, and egg industries. Since the 1930s, when most major nutritional deficiency diseases were brought under control in America, physicians have paid decreasing attention to this field. The teaching of nutrition in medical schools declined and remains at a low level despite the convincing linkages between diet and several important diseases. Many interns, residents, and attending physicians in hospitals came to rely on dieticians to devise their patients' diets, while the doctors themselves continued to consume hospital cafeteria food—the same food which, when experimentally fed to laboratory monkeys, led to the development of severe atherosclerosis.

By the early 1960s the field of nutrition was dominated by Basic Four types who saw nothing but good in the foods they recommended, even though nearly a million people a year were dying from cardiovascular disease. At the same time, American research scientists were rediscovering what the Russians had discovered fifty years earlier.

One of the earliest findings to surface was that American men had a sixfold greater chance of dying of a heart attack than Japanese men. Further investigation of this led to some startling facts.

Using careful analytic methods, Dr. M. G. Marmot of the University of California at Berkeley and his colleagues ran a 1975 study of Japanese men in Japan and those who had emigrated to Hawaii and California. They found that those who had emigrated to California ate more saturated fat and cholesterol than their counterparts in Japan. Their death rate from coronary heart disease was 3.7 per 1,000, as compared to 1.3 per 1,000 for the Japanese men in Japan. The Japanese in Hawaii had a dietary pattern intermediate between that of the subjects in Japan and those in California, and their death rate from coronary heart disease was 2.2 per 1,000.

These findings confirmed what had been demonstrated in the International Atherosclerosis Study in the 1960s—that both the dietary intake of saturated fat and blood cholesterol levels correlated positively with the severity of coronary atherosclerosis. This study clearly showed that life-style and diet, rather than heredity, primarily determine the development of atherosclerosis. (If atherosclerosis were hereditary, one would have expected the three groups of Japanese to have had the same death rate from coronary heart disease.) Other studies, of Italian and Irish immigrants, have confirmed that coronary heart disease is first and foremost related to diet. Immigrants assume the risk of the country to which they migrate because they assume its dietary patterns.

In 1970 Dr. Ancel Keys of the University of Minnesota School of Public Health and his colleagues published the results of a seven-country study that yielded amazing data on the pivotal role of diet—and the additional roles of hypertension and cigarette smoking—in coronary heart disease. The study was conducted among twelve thousand men aged forty to fifty-nine in Finland, Greece, Italy, Japan, the Netherlands, the United States, and Yugoslavia. The highest incidence rates for coronary heart disease over a five- to ten-year period were found in the United States and eastern Finland. This positively correlated with the levels of saturated fat ingestion. The intake of fat correlated with cholesterol levels in the blood, and both correlated with the occurrence of coronary heart disease five and ten years later.

These epidemiologic studies have been complemented by experiments in animals. Monkeys fed diets high in fats and cholesterol for long periods develop advanced atherosclerosis. In some experiments, monkeys fed diets straight out of university hospital kitchens have developed severe atherosclerosis.

The evidence linking saturated fat and cholesterol intake to coronary heart disease continues to mount, but it has not convinced everyone, some scientists included. Several epidemiologic studies have been unable to show a significant statistical correlation between nutritional components in the diet and the blood cholesterol levels of individual subjects. Most scientists not connected to the food industry believe that this lack of correlation is due to problems in the design and methods of such studies. The scientific allies of the meat, egg, and dairy industries, however, have seized upon this as a reason for not recommending a reduced dietary intake of cholesterol and saturated fat for adults who are not overweight, and the mere existence of such studies, no matter how flawed, gave members of the Food and Nutrition Board of the National Research Council the scientific basis for a desperate effort on behalf of eggs, dairy products, and meat. In May 1980 they made front-page headlines by stating that they found *no reason* for healthy Americans to restrict their consumption of cholesterol. The Board further said that it found no clear-cut evidence that reducing the levels of cholesterol in the blood could prevent coronary artery disease.

The Board's pronouncement not only produced headlines— it raised a decidedly mixed reaction in the scientific community, and sowed confusion among average Americans. Farmers and food industry executives—personal friends of many members of the Board and financial supporters of their research— were delighted with the announcement. The American Heart Association was not, charging that the Board had ignored a vast body of scientific evidence. *The New York Times* characterized the Board as "a group whose objectivity and aptitude is in doubt."

The surgeon general and other government health officials stood by their advice—eat less fat and cholesterol. Yet the American Medical Association, which quietly advocates a diet low in fat and cholesterol, but which has never made an explicit statement about the relationship between cholesterol and coronary heart disease, did not condemn the report. This very well served the interests of the National Dairy Council, the United Egg Producers, and the National Livestock and Meat Board, whose advertising and public education efforts have been directed at getting people to eat foods high in fats and cholesterol.

The average person confronting these conflicting views was understandably confused. And this confusion led many people who had followed the controversy to continue habits not far different from those who were totally uninformed.

Another confusing issue is the suspected link between high-fat diets and breast, lung, prostate, and (especially) colon cancer. At the very end of 1983 the American Cancer Society said that recent research findings suggested that high-fat foods did, indeed, increase the risk of cancer. The Society suggested cutting consumption of saturated and unsaturated fats by 25 percent and increasing intake of certain fruits, vegetables, and whole-grain cereal products, especially those high in vitamin C and beta carotene (which is converted by the body into vitamin A).

In the March 8 and March 15, 1984, issues of the *New England Journal of Medicine,* Dr. Walter C. Willett and Dr. Brian MacMahon of Harvard University's School of Public Health reviewed the scientific literature on the subject and essentially agreed with the American Cancer Society's recommendations. However, they pointed out that information about specific dietary factors and cancer is inconsistent and incomplete, and that the data on hand are not sufficient to allow for "strong specific dietary recommendations."

As a rule, scientists and medical researchers make poor players in the complex game of special-interest politics—although they often think otherwise. They are not well-endowed with the stamina, patience, and shrewdness that this game requires, and deep down they view it as an anti-intellectual activity beneath their scholarly dignity. Even when organized into illustrious professional groups they shrink from combat and bloodletting. This is more a reflection of the unsuitedness of their training and temperament to the political arena than it is a mark of weakness of conviction. It was therefore no surprise that when more experienced players from the meat, egg, and dairy lobbies rushed out to do battle, the scientists retreated from the field.

A dramatic example of this occurred in 1970 when dairy farmers made a concerted effort to silence the American Heart Association.

In the 1960s the AHA funded a great deal of research into the connection between high-fat diets and coronary heart disease, and led the public battle against high-fat diets. Without much help from any other voluntary or government groups, the Association almost singlehandedly made cholesterol and saturated fat major public health issues. Most other voluntary, scientific, and governmental groups concerned with public health remained conspicuously silent, preferring to stay out of what they perceived as a brewing political battle of enormous dimensions. And when the going got very rough, even the American Heart Association fell silent.

First, the dairy farmers threatened the AHA with a multi-million-dollar lawsuit for disseminating what they called misleading advice. Their threat had no great effect on the national AHA, but when directed at the local Wisconsin chapter it brought about a marked change in public policy. The Wisconsin Heart Association was told by dairy farmers that if it continued to support the dietary advice of the national organization it would find it hard to raise money in that dairy state.

Frightened by the prospects of a costly lawsuit and dwindling donations, the Wisconsin Heart Association capitulated. It was all done with appropriate window dressing, including the formation of a special Task Force on Nutrition and Cardiovascular Disease. This task force, whose members included the executive director of the Dairy Council of Wisconsin, heard testimony from all sides and finally recommended a new position that was quickly adopted by the Wisconsin Heart Association. In one press release, the Association said, "If hypercholesterolemia [high blood cholesterol] is not identified as a risk factor in an individual . . . Wisconsin Heart does not recommend a low-cholesterol, low-fat diet to prevent heart attacks and strokes." Essentially, the Association recommended dietary changes only *after* high blood cholesterol levels had developed. This new position was praised by dairy cooperatives throughout the state and by the Dairy Council of Wisconsin. More importantly, both the cooperatives and the council dropped their threats of a lawsuit.

The AHA was shocked by the events in Wisconsin, but was powerless to do anything about them. The Wisconsin group was told that it could not take public policy positions contrary to those adopted by the Association's national board of directors, but the local group simply shrugged and went its own way, deferring to its instincts for survival.

While the steady buildup of scientific data incriminating meat, eggs, and dairy products in the twentieth-century epidemic of coronary heart disease eventually mandated government intervention, that intervention was slow in coming and faltering when it did arrive. The battle lines were drawn between a formidable alliance of meat, egg, and milk producers (and their political and scientific allies) and a much weaker assemblage of independent medical researchers, public interest and consumer groups, and a handful of political leaders.

The proponents of saturated fat and cholesterol had much in their favor. Representing important segments of the national economy, they had decades of congressional lobbying experience. They had also been actively supporting nutrition researchers for years, and could easily muster an array of scientific opinion to refute the claims of medical researchers who they saw as newcomers and interlopers. At the grass-roots level, cattlemen, dairy farmers, and egg producers held the keys to the election and reelection of congressmen and senators in a number of states. And finally—and perhaps most importantly— these industries had nutrition-education track records that stretched over decades; at their disposal were sophisticated and well-established national nonprofit organizations which represented their interests and enjoyed public confidence.

Those fragmented groups and individuals who chose to do battle with this powerful consortium were handicapped by passivity on the part of organized medicine, and less than enthusiastic support from those governmental bodies that should have been in the front lines.

There were many reasons for this. For years the National Heart, Lung and Blood Institute, one of the most important divisions of the National Institutes of Health—which has still not made public recommendations about dietary cholesterol and saturated fat—cited a lack of causal evidence as the reason. But Patricia Hausman, of the Center for Science in the Public Interest, who meticulously studied the history of the Institute's policies concerning dietary fat and cholesterol, has unearthed other reasons. In her book *Jack Sprat's Legacy*, she states that the Institute stalled its prevention efforts out of fear of losing funds for research, its most important activity. (In other words, research scientists were supposedly afraid to admit that enough information was in hand on the cholesterol question, because if they did, Congress might reduce their budgets and shift the monies to nutrition education programs.) She also believes that the Institute feared reprisals from farm groups in the form of lawsuits, negative publicity, and, most importantly, lobbying efforts in Congress that might have led to a reduction in its budget.

The U.S. Department of Agriculture, the federal agency responsible for nutrition education, is hardly likely to take the lead in a battle against its own constituents, the meat, dairy, and egg industries. Department directors are often ex-farmers themselves, and some, on leaving office, have assumed lucrative positions with organized farm interest groups. Then too, departmental officials, like the farmers themselves and the Basic Four nutritionists, are convinced of the validity of their position. Many of them practice what they preach, eating lots of eggs, meat, and dairy products and encouraging their families to do the same. Nutritionists who have been advising two generations of parents and children to eat eggs, drink plenty of whole milk, and eat prime meat to avoid beriberi, pellagra, and other vitamin-deficiency diseases, can scarcely be expected to admit that they have been giving bad advice.

In fact, it is unlikely that most of these people will ever admit this. Max Planck, the German physicist, once observed that a generation usually does not change its views radically. New ideas take root only because the older ones die out with the generation that held them.

And so those who choose to take on the saturated fat and cholesterol lobby confront not only powerful economic interests fighting for survival, and politicians who depend upon the farm vote, but also an array of nutritionists and scientists who believe that the new evidence is unconvincing. They are sincere, but wrong. Unlike the fifteenth-century geographers who steadfastly held that the earth was flat or was the center of the solar system, these true believers, by advocating their views, are killing hundreds of thousands of people, and are likely to continue doing so for some time to come.

In contrast to their opponents on this issue, the farm lobbies that represent the big three industries are numerous and powerful.

The National Live Stock and Meat Board, based in Chicago, receives about $5 million annually from livestock producers and meat processors. This money is used to fund nutrition research, prepare recipes for newspapers, and design nutrition education programs in schools. These efforts are geared to defending meat by admitting that cholesterol is a risk factor for coronary heart disease while avoiding the issue of meat's saturated fat content. (This strategy recognizes the now widespread awareness of the link between dietary cholesterol and heart disease, but takes advantage of public ignorance that the body's chemical processes convert dietary saturated fat, which is abundant in meat, into cholesterol.) The board also publishes carefully selected pieces in its newsletter, *Food and Nutrition News*, that support its position while simultaneously ignoring information that is harmful to it. And it has at its disposal a number of Basic Four–oriented nutritionists who willingly give credence to its position.

Complementing the activities of the board are those of the National Cattlemen's Association, a Washington-based lobbying group that represents the country's 1.3 million cattlemen. This group, working in concert with others, has had a long record of successes in bringing about legislation favorable to the meat industry. In 1977 this group, along with the Meat Board, was able to alter the wording of the second edition of *Dietary Goals for the United States,* a report by the Senate Select Committee on Nutrition and Human Needs, so that the phrase "decrease consumption of meat and increase consumption of

poultry and fish" in the report was changed to "decrease consumption of animal fat, and choose meats, poultry, and fish
which will reduce saturated fat intake." The two organizations
succeeded in this largely through pressure on Senator Robert
Dole of Kansas, a prominent farm-state member of the committee. But, not especially happy even with the new wording,
the Cattlemen's Association and the Meat Board then denounced the whole report as inaccurate and misleading.

Like the cattlemen, the dairymen are well organized. They
number only four hundred thousand, but have had an even
greater impact on American eating habits than the meat groups.
Their Washington-based National Milk Producers Federation is
a powerful congressional lobby that has used its muscle to prevent government agencies from informing Americans about the
link between diet and coronary heart disease.

The Federation was largely instrumental in having the Senate
Select Committee on Nutrition and Human Needs cancel its
1971 public hearings on diet and heart disease. This cancellation was sharply attacked by Dr. Jean Mayer, America's most
respected nutritionist and now the president of Tufts University, who wrote in a letter to The Washington Post that the
pressure of dairy farmers on captive politicians had prevented
an examination of the most important health problem facing
the nation. Eventually the hearings were held, but it was not
until 1977, six years later, that the final Dietary Goals report
was published.

Political Action Committees (PACs) for the dairy industry
exert enormous influence on legislators. "The dairy industry
spreads an awful lot of money around," says Robert Michel, an
Illinois Republican and House Minority Leader, "and that gets
reflected in votes out here, I'm afraid." In October 1981, for
example, the House of Representatives defeated, by a vote of
243 to 153, an amendment that would have reduced dairy price-
support subsidies. Common Cause magazine later reported that
the three dairy PACs gave the 243 members who voted to defeat
the proposal a total of $1 million in campaign contributions
during the 1978 and 1980 elections. This amounted to an average of $4,271 per representative, versus the $718 per representative (and total of $109,900) for the 153 representatives who
voted against the dairy industry's position.

The American Dairy Association, meanwhile, works the TV screen and roadside billboards for the industry, pushing milk, cheese, and butter. In recent years it has used catchy jingles and slogans like "Milk, the Fresher Refresher," on a budget of $3.5 million a year, which is donated by dairy farmers.

The National Dairy Council's main goal is to carry the dairy producers' message into schools and newspapers. It has had an immense role in shaping the eating habits of two generations of Americans, and its nutrition education materials are still widely used in school districts throughout the country. Indeed, even though Congress has since 1973 allocated from $15 to $30 million per year for a National Nutrition Education Training Program aimed at teaching students, children, and school cafeteria personnel about nutrition, twenty-nine states use this money to buy the Dairy Council's messages.

Emphasizing the nutrient value of dairy products, the NDC's publications fail to say anything about their high saturated fat content. Nor do they mention that blood cholesterol levels are greatly influenced by saturated fat in the diet.

Bolstering the position of the Dairy Council and dairy farmers in general are government policies that foster a high saturated fat intake.

In most states, for example, milk is priced on the basis of its fat content—a system that Dr. Stephen Babcock of the College of Agriculture of the University of Wisconsin developed in 1890, when the illegal watering and skimming of milk was rampant. Over the past eight decades, this pricing policy has led to the breeding of cows that produce milk with a high fat content. Today, many cows whose low-fat milk is far better for consumers are sent off to the slaughterhouse. Furthermore, the government not only buys all the surplus milk in the country, distributing it in the form of butter, cheese, and powder to prisons and schools, but it also has enacted laws that discourage the use of margarine in restaurants and the marketing of cheeses in which corn oil replaces saturated fat. Most such actions have come about through pressure from the National Milk Producers Federation.

While meat and dairy interests have made costly offensives their best defense, this has not worked well for the United Egg

Producers or the National Commission on Egg Nutrition. The former is a Washington-based lobby group, the latter a group established in 1971 to fight the American Heart Association's stand on dietary cholesterol and heart disease.

Egg consumption in the United States has been gradually falling for several years, not necessarily because people have heeded the advice of the American Heart Association, but because sit-down breakfasts have become rare in American homes. So great was the fall in egg consumption—30 percent in thirty years—that in the early 1970s the National Commission on Egg Nutrition launched an advertising campaign that eventually brought it to grief with the Federal Trade Commission. The ads claimed, among other things, that cholesterol is the richest source of protein in human nutrition, that it is the building block of sex hormones, that it is required for proper functioning of the body's nerves, and that the body normally eliminates the same amount of cholesterol as is eaten, while producing more cholesterol the less cholesterol one eats. The American Heart Association brought a complaint before the Federal Trade Commission, claiming that these ads contained incorrect and deceptive information. Eventually the National Commission on Egg Nutrition had to withdraw the ads.

But the egg industry did not stop with this defeat. They went the route of Congressional lobbying and were successful in having the Egg Research and Promotion Act passed. This act set up a national organization to promote eggs, funded with taxes on egg producers and administered by the U.S. Department of Agriculture. To disarm the American Heart Association and medical researchers, the act provided for generous research monies to be held out as tempting bait to the egg industry's opposition. Some heart researchers, hard pressed for funds, were only too happy to accept the egg money, and some of them eventually went on to produce research results showing that eggs had no effect on cholesterol in the blood.

Three of these studies were conducted at the University of Wisconsin, the University of California at Los Angeles, and the University of Illinois. The general public may have been impressed that these prestigious universities were linked with data that egg intake doesn't affect blood cholesterol. However, the National Heart, Lung and Blood Institute (NHLBI), after a careful examination of these research results and the studies that led to them, pronounced them full of flaws and concluded that they were meaningless.

These results nevertheless provided the egg industry with what it wanted—the means to say that "eggs don't raise cholesterol." And it promptly did so on TV, on radio, and through flyers inserted in egg cartons. Few consumers could judge who was right, the NHLBI or the egg industry's scientists.

In its *Dietary Goals for the United States*, the Senate Select Committee on Nutrition and Human Needs challenged the cozy relationships that exist between government, the meat, dairy, and egg industries and the Basic Four nutritionists. Following thousands of hours of expert testimony, the report advocated big reductions in total dietary fat, saturated fat, cholesterol, and sugar.

The prompt attack on the study by the meat, dairy, and egg lobbies and their scientific and political allies made it appear "controversial" to the general public, which is what the industry strategists wanted. They were overjoyed with *The Washington Post's* headline: "Tug-of-War Over Diet: Nutrition in America Becomes a Political Hot Potato."

Overriding industry opposition, the consumer-oriented Carter administration pushed through several government reports on diet that had been stalled for almost a decade. On July 28, 1979, Joseph Califano, Secretary of Health, Education and Welfare, and Dr. Julius Richmond, Assistant Secretary of Health and Surgeon General, released a document entitled "Healthy People: The Surgeon General's Report on Health Promotion and Disease Prevention," in which they advised people to eat less saturated fat and cholesterol, salt, and sugar. In it, they said, "A good case can be made for the role of [a] high intake of cholesterol and saturated fat, usually of animal origin, in producing [the] high blood cholesterol levels . . . associated with atherosclerosis and cardiovascular diseases." This was a very powerful statement, coming as it did from the Surgeon General and the Secretary of HEW. Soon afterwards the National Cancer Institute came out with similar recommendations.

Even the position of the Department of Agriculture changed, largely because Secretary Robert Bergland and Assistant Secretary Carol Foreman were able to rid the department of much of its control by the farm industry. In what represented a revolutionary departure from previous policies, the department copublished, in February 1980, a booklet that advised people to reduce their dietary intake of fat and cholesterol. Entitled "Nutrition and Your Health: Dietary Guidelines for Americans," it was fiercely attacked by the farm lobbies and by farmbelt members of Congress.

The meat, dairy, and egg industries continued to do battle, losing ground most of the way. Then, with the election of President Ronald Reagan, their fortunes began to turn. In 1982 the Agriculture Department's magazine *Food/2* was about to publish an article advocating weight control and a modified fat and cholesterol diet. The farm lobbies, enraged, pressured the department to squelch the publication. Secretary of Agriculture John R. Block (himself a farmer) announced that the magazine would be published, but *without* the chapter on fat and cholesterol.

Entitled "Eating the Moderate Fat and Cholesterol Way," this excised chapter made rather mild statements about fat and cholesterol. It also acknowledged that there were still doubts about the desirability of making general recommendations about dietary fat and cholesterol to the entire American population— although it did say that the Surgeon General and others felt that it was sensible to consume only moderate amounts of these substances.

But the meat, egg, and dairy industries wanted nothing of even mild criticism—of any of their products. They increased their pressure on Secretary Block, who finally capitulated and announced that the magazine would not be published at all. (Deputy Secretary Richard E. Lyng, former president of the American Meat Institute, had said publicly that the magazine would be "published over my dead body.") The Department of Agriculture had once again become a fiefdom of the farm interests.

To appease consumers, the Department of Agriculture asked the American Dietetic Association to publish the magazine without the fat and cholesterol chapter—but the Association refused. Instead, the ADA opted to publish the magazine's three chapters as three separate booklets.

The New York Times described Secretary Block's decision with the headline "CONTROVERSIAL DIET TO BE PUBLISHED," and similar headlines appeared all over the country, confounding and dismaying millions. By creating the illusion of controversy, the saturated fat and cholesterol industries continue to confuse and misinform the American public.

In December 1982 the Department of Agriculture moved to set up a panel to review its dietary guidelines. Nine scientists were appointed to the panel, six by the USDA and three by the Department of Health and Human Services. Several of the USDA's appointees had known ties to the dairy, meat, egg, and sugar industries. In making these appointments, the USDA didn't even try to give the appearance that the panel was objective.

While the Reagan Administration was successful in promoting dairy, meat, and egg interests through tampering with the dietary guidelines, the Federal Trade Commission was taking some small steps in the opposite direction. The FTC began a survey of advertising claims made for foods containing fat and cholesterol as a first step toward regulating such advertising. Michael Pertschuk, a former FTC chairman who had pushed for regulation in this area, commented that "It may well be a cornerstone of a free market that consumers have an inalienable right to clog up their arteries if they want to do so," but, he added, "There are two modest proposals that should go along with that freedom. One is that consumers, particularly those looking for healthful alternatives, not be misled by deceptive health claims for food. The other is that consumers should be informed about the risks associated with those choices." Many doctors think that the labels on processed foods should reveal the amounts of cholesterol and fats the product contains. Even such a simple substance as a soda cracker contains four grams of saturated fat, and saturated fat, usually in the form of coconut oil, is found in most commercially baked breads and cakes.

But now the results of what has been widely described as a "landmark" study may have tipped the scales in favor of cholesterol control forever. Government-supported researchers taking part in the ten-year project announced in January 1984 that they had produced "the first study to demonstrate conclusively" that the risk of coronary heart disease can be reduced by lowering blood cholesterol. The results of their multicenter research effort were published in the January 20, 1984, issue of the *Journal of the American Medical Association*. Previous studies had *associated* high blood cholesterol with cardiovascular disease, but whether cholesterol reduction could actually reduce heart disease had remained an open question. The study, involving 3,806 men, and supervised by the National Heart, Lung and Blood Institute, found that using the potent cholesterol-lowering drug cholestyramine produced the greatest effects (a 24 percent reduction in fatal heart attacks), but that changes in diet also proved positive.

George Lundberg, editor of the *Journal of the American Medical Association*, predicted that twenty-five years from now this would be looked upon as the study "that secured the cholesterol theory of coronary heart disease." Dr. Robert I. Levy, vice president for health sciences at Columbia University and a longtime cholesterol researcher, was even more adamant. "With the validation of current hypotheses . . . the current epidemic of cardiovascular disease," Levy said, "should go the way of the epidemics of infectious diseases that besieged the country earlier this century."

But that day still seems a long way off. In the wake of the NHLBI report the food industry was unrepentant. "Most of us can eat one or two eggs a day without problems," said Louis Raffel, president of the American Egg Board. And Dr. Edward Ahrens of Rockefeller University said: "To deny everyone red meat could mean taking away the joy of life unnecessarily. And as an inexpensive source of good nutrition, there is nothing more glorious than the egg."

3
Imposed Risks

While many risks are chosen, others are imposed. Some of these risks, as we have seen, are chosen by one individual and then passed on directly, or indirectly, to others. The average "problom drinker" in the United States, for example, costs society as a whole roughly $10,000 (in health care and other expenses) every year. But many other risks are created not by individuals, but by technological and industrial processes, products of dubious safety, and chemicals that find their way into our air, food, and water.

As with chosen risks, the reasons for our exposure to imposed risks are not clear-cut. We could (or should) know about some of the hazards imposed on us, and take evasive action, but often we do not. In most cases, however, industry or our government watchdogs could or should have prevented the exposure in the first place.

As we have noted, people seem to make a strong distinction between risks they themselves choose and those that are imposed on them. "What makes people understandably angry," Mary Douglas, a social anthropologist, and Aaron Wildavsky, a political scientist, have written, in their controversial book *Risk and Culture*, "is damage that they feel they should have been warned against, that they might have avoided had they known, damage caused by other people, particularly people profiting from their innocence." This, Douglas and Wildavsky argue, "corresponds closely to legal conceptions of employers' liability and the responsibility of producers to sell safe goods." In other words, the fact that we freely embrace certain risks does not give industry or the government the right to impose similar or quite different risks on us.

The range of imposed risks is frightening. However, just as there are few "pure" chosen risks (such as sky-diving), so too are there few risks that are absolutely imposed: a sudden thunderstorm, some forms of cancer, or a madman with a rifle loose in the community are examples of these.

One week after the May 18, 1980, eruption of Mount St. Helens in Washington, which took thirty-six lives, *The New York Times* termed it a tragedy with "no guilt." The newspaper said that there was "nothing anyone should or could have done. You can't blame a volcano."

The volcano remains blameless, but in the months following the eruption some people began accusing the media of not raising questions, before the tragedy, about the so-called "red zones"—areas that should have been off-limits to the public. And relatives of some of the victims later sued the state of Washington and the Weyerhaeuser Company (a huge lumber concern) charging that they had negligently allowed loggers to work too close to the volcano's summit when an eruption seemed imminent.

Every act of nature is "imposed" on us, but we often leave ourselves open to that risk, such as by living in an earthquake- or flood-prone area. Frequently we meet imposed risks halfway. Noise pollution has become an almost unavoidable headache in modern society; each day approximately twenty million Americans are exposed to noise levels that can permanently damage their hearing, according to the EPA. A recent survey at the University of Tennessee found that more than 60 percent of that school's freshmen already had significant hearing losses in the high-frequency range. But some of the students had unquestionably brought that condition on themselves by listening for long periods to high-volume rock n' roll.

"Voluntary/involuntary is a movable boundary," Douglas and Wildavsky write. If, like chosen risks, imposed risks are often relative, how do we determine how "angry" to get about them? This may rest on the notion of reasonableness. Getting hit head-on by a car driven by someone who has been drinking heavily is surely an imposed danger—although we *could* choose to avoid it by not driving at all. But is this option reasonable? Most people would say no, and therefore call for a more reasonable alternative: stronger curbs on drunken drivers.

Of course, what is reasonable in one era—such as allowing 18-year-olds to drink (and drive)—may be deemed unreasonable in another. And what one person may judge as reasonable another may not. The vast majority of Americans accept the routine addition of fluoride—a *possibly* harmful chemical—to their drinking water, but some do not. They drink bottled water or try to force their communities to halt fluoridation entirely. (In 1981 forty-one communities around the country held referendums on fluoridation and thirty-three voted to stop it.)

Fluoridation used to be more hotly debated. But since the release in 1963 of the film *Dr. Strangelove*, which featured a General Jack B. Ripper who believed that "fluoride is the most monstrously conceived and dangerous Communist plot we have ever had to face," those who oppose this additive have been widely considered to be lunatics.

Critics claim that fluoridation was foisted on an unsuspecting America in the 1950s after undergoing inadequate health tests. But blessed with claims (disputed by many) that it prevents two out of three cavities, it is now found in the drinking water of six out of ever ten Americans, and, in addition, in 90 percent of all toothpastes sold in this country. While the weight of scientific evidence indicates that the level of fluoride added to water supplies is safe, recent reports provide some food for thought (or at least this issue out of the "crackpot" category).

Some foreign studies show, for example, that human exposure to fluoride from all sources is rising dramatically and already approaches or surpasses safe levels. This means that fluoride levels in drinking water, once safe, may now be dangerously high. A report by the Ministry of the Environment of Quebec in 1979 concluded that "fluorides are highly toxic for humans and a narrow margin separates an 'acceptable level' from a toxic level," and recommended a moratorium on water fluoridation. A recent study by Japanese scientists suggested that fluoride is a potential carcinogen. (Dr. Dean Burk, a biochemist at the National Institute of Cancer for thirty-five years, contends that some 35,000 cancer deaths per year are caused by fluoridation.) Congress has ordered a federal study, with results due in 1988, which may indicate whether there is, after all, any threat posed by fluoridation to our "precious bodily fluids" (as General Ripper put it).

Perhaps the most regrettable type of imposed risk, and the most common, is one that an industry puts on the public without telling them about it, especially when the government lets the industry get away with it. Because most people are quite willing (perhaps too willing) to balance risks against benefits—to be "reasonable" about risk-exposure—deliberate endangerment often makes us angry.

A striking example of this was played out in August 1983. In a precedent-setting move, the Justice Department sued General Motors, charging that the company had sold 1.1 million 1980 X-model cars which it knew had hazardous brake defects; these had caused numerous accidents and at least seventy-one injuries and fifteen deaths. Two days later, a General Accounting Office study concluded that there were "serious problems" in the government's handling of the X-car problem. The report charged that the National Highway Traffic Safety Administration did not follow its own guidelines for investigating safety-related defects, and had shown an "apparent reluctance" to inform the public.

Nevertheless, it is industry that is usually seen as the prime culprit in the imposed-risk business. Perhaps because they are not perceived as being "out for a buck," government watchdogs are often considered stupid, hapless, lazy, or debilitated by bureaucracy—but not villainous. They are faulted for what they do not do, not for what they do. Although the results of inaction can be as grievous as those of an intentional act, they somehow seem less purposeful.

By contrast, the public perceives industry as quite aggressive. And many company officials (as we show throughout this book) do indeed use a variety of sophisticated strategies to justify potentially dangerous products. These include propagandizing the public (and company workers), influencing public policy, controlling the nature and flow of information, wearing down regulatory agencies, minimizing risks, and using diversionary tactics.

Philosophical as it is about risks generally, the public, with great justification, sometimes rebels against an imposed risk. Unfortunately, the results are not always overwhelmingly positive.

When Americans cried out against the haphazard disposal of toxic chemicals in the late 1970s, industry and the government began looking for new sites where these substances could be safely disposed. This has been a nearly fruitless search. Almost every proposal for a badly needed new dumping ground meets local opposition. People want more safe dump sites— but not near them. Officials have even given this syndrome a name: NIMBY—not in my back yard. (On the other hand: Given industry's track record, who would want to live next to a potential Love Canal?)

And what happened in the wake of Three Mile Island and mass demonstrations against unsafe nuclear power plants? National policy shifted to an emphasis on coal-power. This decreased the risk of radiation poisoning (in the event of a meltdown) imposed on the public-at-large—but also meant that accidental death would be specifically "imposed" on an additional number of workers who mine coal, a most hazardous operation. It also aggravated the problem of sulfur dioxide and other coal burn-off pollutants in the air, and contributed to the likelihood of a "greenhouse effect" (see Chapter 7). In choosing coal over nuclear power, society reduced one imposed risk while expanding others.

The disposal of toxic chemicals, the burning of coal, and the consequences of nuclear power-plant meltdowns are often perceived as imposed risks over which the individual has no control. With regard to corporate polluters, however, Ralph Nader argues that much can be done. He points out that people can avoid personal exposure to some chemicals, and that communities can successfully deter chemical pollution caused by local industry. He advocates the federal chartering of companies to make them more accountable, and calls for criminal penalties for corporate irresponsibility.

Perhaps the easiest of these suggestions to implement is the individual avoidance of certain chemicals. Nowhere can this be done more effectively than in choosing what we eat and which drugs we take. Whether we avoid chemical-laced food and risky drugs depends on our knowledge, insight, motivation, and a healthy level of suspicion. Some drug and food manufacturers have a long track record of knowingly marketing risky products and willfully hiding the risks in the process. Consumers might reason that marketing such products would backfire because their risks would sooner or later become known, but this is not so, since the risk probability may be small enough to go unnoticed indefinitely. Manufacturers have also shown that they are not averse to making a short-term profit from a product that may ultimately be banned, perhaps because they have rarely been fined or jailed.

Today, however, prosecutors are increasingly pressing criminal charges against—and calling for jail terms for—officials whose companies harm the public or the environment. After Stefan Golab, a forty-five-year-old Polish immigrant, died of cyanide poisoning, a murder indictment was handed down against five executives of the film recovery company he worked for in Elk Grove Village, Illinois. The officials were accused of failing to take safety measures and engaging in practices that "created a strong probability of death and great bodily harm." (The trial is pending.) A number of victims of handgun violence recently filed lawsuits against companies that make Saturday Night Specials, charging that these cheap weapons are, by their nature, hazardous, and that manufacturers (and sellers, too) should be liable for injuries or deaths caused by them.

In 1983 more than two dozen polluters were convicted on criminal charges brought by the EPA. An aggressive city attorney in Los Angeles, Ira Reiner, has sent several polluters to jail himself. In one of these cases a company that had dumped toxic chemicals into the city sewer system was sentenced to buy a half-page advertisement in *The Wall Street Journal*, in which it had to admit that it had wrongly endangered the public health, and reveal that one of its executives had gone to jail. Reiner explains that he wanted to put a "chill" in boardrooms across the country.

Despite the existence of the Food and Drug Administration and consumer watchdog groups, many risky foods and drugs escape the net of scrutiny and enter the market. Residues from five to six hundred chemicals, for example, probably end up in the meat we eat, with 143 of them, according to the GAO, likely to be present in amounts above the tolerance levels set by the government. Many are suspected of causing cancer. Yet the government monitors only sixty of these residues, and meat inspections declined during the Reagan Administration. Carol Tucker Foreman, who supervised the residue-monitoring program for the Department of Agriculture in the Carter Administration, told Congress in August 1982: "There is a good chance that the American public consumes meat with violative levels of carcinogenic and teratogenic chemical residues with some regularity."

Many Americans also unwittingly consume unsafe or useless drugs. An eleven-year FDA study of seven hundred key ingredients in nonprescription drugs, released in October 1983, found that only one-third had been proven safe and effective. Consumer groups have been trying to get courts to order the agency to take all unproven drug ingredients off the market.

Perhaps the most flagrant example of a drug company foisting an imposed risk on the public occurred in 1960. At that time, there was growing interest among physicians in reducing the levels of blood cholesterol, since it had been linked to coronary artery disease and arteriosclerosis. Thus, when Merrell Laboratories introduced its new cholesterol-lowering drug triparanol, also known as MER/29, doctors greeted it with great enthusiasm. But the drug wasn't long in use before it began causing cataracts. The FDA ordered the drug recalled, and the story might have ended there had it not been for a former Merrell laboratory technician who sent word indirectly to the FDA that she had been involved in research on triparanol. She had quit her job because she had to suppress the results of tests in monkeys that clearly showed ill effects on their eyes.

The FDA's investigation uncovered the unthinkable. The company had conducted laboratory studies of triparanol in conjunction with two other drug companies. All of the studies had revealed that triparanol had cataract-producing properties, yet Merrell marketed the drug anyhow.

Three of Merrell's officials were indicted on criminal charges, but received suspended sentences; an $80,000 fine was imposed on the company by the courts. Later, 1,500 civil suits from around the country eventually cost the company some $50 million. But this was small comfort to the hundreds of people who suffered permanent eye damage.

Perhaps nowhere in the health field have imposed risks been more numerous than in weight-reduction plans and diet pills. In the past twenty years, the FDA and organized medical groups have taken action against many diets devised by quacks, self-appointed experts, headline grabbers, and get-rich-quick promoters. The promise of many of these diets is that those who follow them will "get slim fast." Many of these pseudoscientific diets are presented to the public in popular book form, often published by well-known houses that give them a degree of respectability. And what experts recognize as the authors' thin credentials often come across on book jackets as impressive accomplishments. Often these books become best-sellers overnight because so many people want to believe that they can simultaneously eat as they please and still lose weight.

While some of these diets are amusing and even laughable, others are deadly serious—and possibly also deadly, because, if followed, they can lead to potentially fatal disturbances in the body's chemistry. One of the best remembered of the fad diets is the *Calories Don't Count* diet. This called for the use of safflower oil capsules in conjunction with the diet. The FDA eventually forced these capsules off the market because of false claims made in the book that they produced weight reduction and reduced the risk of heart disease.

In recent years many diets have come under fire by medical and scientific groups as either potentially harmful or for making false claims. Among these have been such favorites as Dr. Atkins' Diet and the Beverly Hills Diet. Although the FDA can prevent the marketing of diet pills, it cannot ban diet books. But informed public opinion can reject the bad advice and false claims made in some of these books.

Diet pills are a big over-the-counter drug business in America. In 1981 ten billion doses of these pills were consumed by some four million Americans, netting billions of dollars for manufacturers. Many of these pills contain a common ingredient, phenylpropanolamine (PPA), which is also found in cold remedies. This is ironic, for while the FDA has moved to prevent the marketing of amphetamines—which are prescription drugs—for weight-control purposes, it has stood by while PPA, a similar drug, is sold over the counter.

The Thompson Medical Company and Smith, Kline and French now produce fifteen different brands of diet pills containing PPA and other ingredients (including caffeine and vitamins). They are called by such names as Dexatrim, Dietac, Hungrex, New Me, Maxi-Slim, Slim-Line, Vita-Slim, and Ayds candy. The major health risks associated with these pills are strokes, high blood pressure, and psychoses. Although the labels on the packages advise people with high blood pressure not to take pills containing PPA, many users may not know they are hypertensive and run the risk of a stroke. Even one pill has been known to cause a stroke in a sensitive individual.

Yet while the National Clearing House for Poison Control Centers annually receives around ten thousand reports of poisonings caused by PPA, and concern about the safety of these pills has been longstanding among physicians, there has been considerable support for PPA in scientific circles. In 1979, for example, a special panel of physicians and scientists appointed by the FDA concluded that pills containing this substance may help people lose weight. The panel also suggested that the FDA let manufacturers double the daily allowable dose of PPA in their pills from 75 to 150 milligrams. The FDA refused to do this and has spent the ensuing years reviewing the panel's suggestions.

To its credit, the FDA-appointed panel did recommend that the labels on PPA pill packages discourage persons being treated for high blood pressure, depression, heart disease, thyroid disease, and diabetes from taking the pills. But what about those who don't read the fine print on obscure labels? Dean Siegal, a company spokesman for Thompson Medical, told *The New York Times* in 1982 that it would insult the intelligence of American consumers to restrict sales of the pills because an irresponsible minority chose to ignore reasonable warnings. But this reply evades the issue of people who don't know they have a medical condition that would endanger them if they took the pills.

In fact, although pills containing PPA have been marketed since the early 1920s, the FDA has never officially declared them "safe and effective." Nor are medical scientists sure how PPA works in suppressing the appetite. These facts, coupled with the known adverse side-effects of PPA diet pills, make them imposed risks, say many observers, even though dieters voluntarily gobbled them up. In April 1981 Dr. Alan Blum, writing in the *Journal of the American Medical Association* about Wisconsin's successful physician-led effort to thwart the diversion of amphetamines for street sale, characterized PPA as an over-the-counter amphetamine. He called for a ban on the manufacture, sale, and distribution of all products containing PPA. Later that year the Center for Science in the Public Interest petitioned the FDA to ban all dietary aids containing PPA. It is unlikely, however, that the FDA, whose policies have been influenced by deregulation sentiment, will vigorously move against PPA.

Thus, for the foreseeable future, PPA-rich pills are likely to remain on the pharmacy shelf because they are wanted by a vocal segment of the population, contribute significantly to the national economy, and are thought by some scientists to be effective.

The PPA example represents the "soft" end of the imposed-risk spectrum. People can choose what to read, what diet to follow, and what diet pills to take. But they have little or nothing to say about the solutions that drip into their veins when they are hospitalized. During the early 1970s, Abbott Laboratories-Abbott Pharmaceuticals of Chicago was forced to recall close to $43 million worth of intravenous solutions that had become contaminated with life-threatening, disease-causing bacteria.

The problem surfaced simultaneously in a number of hospitals in which patients inexplicably developed bacteremia—a serious and often fatal disease caused by bacteria circulating in the blood. At first, physicians treating these patients were perplexed because they couldn't find any source for the infections; quite often the patients had been hospitalized for noninfectious problems such as heart attacks and bad backs. But as epidemiologists from the Centers for Disease Control probed further at two hospitals, they found that the one thing common to all of the infected patients was that they had received Abbott intravenous solutions. Eventually, a total of 378 cases of bacteremia traceable to the solutions were identified in twenty-five hospitals around the country. Some forty patients died.

These findings all pointed to contamination at the place where the solutions had been manufactured. The Centers For Disease Control's epidemiologists quickly learned that Abbott, shortly before the epidemic had begun, had produced a new lid for its intravenous solutions that contained a plastic liner; previously, the company had used a shellacked inner lid. Understandably, the epidemiologists focused attention on the new lids. Sampling 1,007 of them, they found that nearly forty percent were contaminated with bacteria, of which 8.1 percent contained the two organisms that had caused the cases of bacteremia. Further research revealed that although the lids had been contaminated at the manufacturing plant, where the organisms were present in numerous places, the solutions were still sterile on arriving at the hospitals where they were to be used. The inner lids contaminated the solutions when medications were added and the bottles were shaken to insure mixing. The shaking released the bacteria from the lids and, once in solution, they grew rapidly in large numbers.

A number of patients and their families sued Abbott, but these suits all resulted in mistrials. Finally, Ralph Nader's Health Research Group pressured the federal government to take action against Abbott. However, the government's case was undercut from the outset, when it sought an indictment against Abbott because of contaminated solutions rather than contaminated lids. Abbott lawyers were able to successfully convince the court that the lids were at fault, which they were. Because of this, the court fined Abbott only $1,000. Abbott executives felt that this vindicated the company of any wrongdoing. The company had not intentionally marketed a contaminated product. A new technology (plastic liners) had given rise to a new, unforeseen risk.

Imposed risks such as these are but a small sampling of those to which modern man is exposed. One of the most insidious, in fact, is loss of employment, or poverty. According to one recent study, a 10 percent increase in business failures results in almost 6,000 additional deaths from cardiovascular disease. During recessions smoking increases, and so does the overall mortality rate. When people lose their jobs they often lose medical insurance benefits, and their health care frequently suffers. So does their psychological state. Each percentage increase in the unemployment rate is associated with 318 additional suicides, a 5 to 6 percent increase in homicides, and a 4 to 5 percent rise in first-time admissions to mental hospitals.

Most persons lack the knowledge and experience needed to guide them in facing risks, and their resolve to take action is often weakened by a feeling of impotence. Others may be willing to accept an imposed risk because they think that it will bring a benefit that outweighs a perceived low-risk probability. The EPA recently found, for example, that catalytic converters, which screen out pollutants in unleaded gas and are now required on new cars, had been removed from 4.4 percent of the automobiles they checked, and ruined by the unprescribed use of leaded gas on another 13.5 percent. This means that almost one in five car owners are taking steps to save themselves a few cents per trip (leaded gas is cheaper than unleaded) while increasing the risk of lead toxicity for the rest of us.

Government regulatory agencies have historically followed the path of least assistance in confronting major hazards. State officials in California cannot do much to prevent a major earthquake, for example, but this is no excuse for the pathetic level of preparation for the one that will soon hit that state. Some problems must reach crisis proportions and embarrassing public headlines before the bureaucracy will act. How often are we presented on the evening news and in the newspapers with stories involving imposed risks? The human-interest tragedies of innocent victims are graphically described, as well as the seeming irresponsive behavior of government officials. The scenario that then unfolds is almost predictable. Suddenly the government agency swings into action, not so much because it wants to, but because it now has to. It coolly deflects industry criticism with the argument that the company's very survival and credibility are at stake! Unless the imposed risk is clearly an immediate threat to life, which is often not the case, the process of inquiry often becomes an end in itself.

Making decisions is difficult, but disseminating information is not. While they conduct tests, devise cost-benefit formulas, or simply procrastinate, our government watchdogs could at least give us a fair idea of the risk imposed by a particular threat. Information may not be the remedy for an involuntary risk, but it at least tells us that a certain sickness is spreading. Forewarned, "people will either refuse a known risk or seek additional compensation for assuming it," say Mary Douglas and Aaron Wildavsky. Thus, they contend, knowledge converts an involuntary risk due to ignorance into one that is averted—or into one accepted voluntarily.

The limits of this approach, however, are suggested by the following. There are probably thirty-one thousand schools in the United States which house unstable, fiber-shedding asbestos, which causes cancer and other serious lung diseases. These schools are attended by over fifteen million children. The federal government has not taken steps to clean up the schools itself because each renovation costs about $100,000, but it does require local officials to conduct asbestos inspections and tell parents and teachers what they find. Then it's up to the local school districts to find the money to carry out their own remedial action. Not many have done so. Few have $100,000 on hand, and local taxpayers have not often offered to pay the price.

Does this information exchange make asbestos poisoning a "voluntary" risk for schoolchildren? And does it really take the federal government off the hook? Only agencies in Washington have budgets large enough to launch a crash clean-up program. Local school officials are so fearful of what they might find that at least one-third of the schools are not even meeting the minimal inspection requirements.

Imposed risks continue to proliferate. Among those that have surfaced in the past few years are acid rain, toxic shock syndrome, EDB, "indoor pollution," AIDS, toxic wastes, and low-frequency electromagnetic radiation. Because of technological progress we are living more dangerously than ever. For all their benefits, for example, computers have the potential to go disastrously haywire, through malfunction or manipulation. In 1984 a password that could permit access to the credit histories of 90 million people was stolen from a Sears, Roebuck store and posted on an electronic "bulletin board." The password could have been used for up to a month to obtain credit card numbers and charge merchandise, according to TRW Information Systems, the nation's largest credit reporting company. Some computer experts have suggested that enemies abroad could carry out an electronic funds transfer that would drain money out of the United States and destroy the economy. A terrorist attack carried out via computer could severely disrupt transportation and communications across the country; there have been at least thirty attacks on computer facilities in Europe since 1978, in some cases with explosives. An attack on Italy's Motor Vehicle Ministry was so successful that for two years the government did not know who owned vehicles or had a driver's license. The scenario for accidental nuclear war caused by computer breakdown or break-in is well known. Receiving scant attention, however, is the fact that weapons systems of all kinds are rapidly becoming computerized. In July 1983 a computer malfunction on a U.S. warship off the coast of California caused a shell to be fired in the wrong direction—toward a Mexican freighter.

Yet today rules and laws to prevent the misuse of computers are still vague or nonexistent. And some computer experts argue that it is impossible to build safeguards into computers that will prevent any risk of catastrophic accidents.

Everyone recognizes another modern threat—nuclear war— as a major risk. But few have considered the *fear* of nuclear war as a risk in itself, and a potentially far-reaching one at that. Although results fluctuate from year to year, public opinion polls have long shown that a majority of Americans feel that nuclear war is likely by the end of the century. This perceived threat, contends Joel B. Stemrod, an economist at the University of Minnesota, has had "a significant impact on people's willingness to postpone present consumption in favor of investment," and thus contributed to the slowdown of capital accumulation over the past thirty-five years. Fear of nuclear war has reduced the normal inclination to save for retirement, or put away money for one's children. After conducting a study for the National Bureau of Economic Research, Stemrod concluded that that those who disregard this "may be ignoring a significant factor in the performance of the U.S. economy since the beginning of the nuclear age."

But some progress has been made in tackling the sources of many risks. "We are putting firmer reins on some technologies, such as nuclear reactors, whose technical development simply outruns society's management," says William W. Lowrance, a leading authority on risks and director of the Life Sciences and Public Policy Program at Rockefeller University. And, he adds, "we are cleaning up some messy legacies, such as those of asbestos and toxic waste, about which both our scientific understanding and our values have changed. We have become more restrained in releasing materials into the environment. And we have become more cautious about taking irreversible actions."

Lowrance believes that we now experience an awareness of risks that is "tragic" because we know we must make difficult choices to accept some risks and attack others. "In our knowing so much more and aspiring to so much more," he says, "we have passed beyond the sheltering blissfulness of ignorance and risk-enduring resignation." This, Lowrance observes, has generated a good deal of apprehension among both individuals and institutions. "Similar risk-changes," he says, "have occurred historically when people became aware of specific causes of disease and deformity, as when it became clear that moral turpitude alone was not the cause of syphilis." But, he concludes, the difference today is that though it is discomfiting, "the conjunction of greatly improved science with heightened societal aspiration now puts us in a position to manage aspects of the 'tragedy' as it unfolds."

Close-Up (1)

ORAFLEX

"The Oraflex case is representative of a widespread problem which continues to grow because there has never been any prosecution of companies failing to promptly and completely report adverse reactions to drugs, before or after marketing."
—Dr. Sidney Wolfe,
Health Research Group

There are close to thirty-one million arthritis victims in the United States, whose joints ache and swell and are sometimes destroyed to the point of severe deformity and disability. Many of these people live with pain night and day. For some, aspirin—which is still the single most effective drug for relieving the pain of arthritis—affords temporary relief. But for others, aspirin does little to relieve their grinding pain.

Patients and doctors have long hoped for a drug that would provide lasting relief for both the inflammation and pain of arthritis. Every prospective drug that has come along, like cortisone and chloroquine (the antimalaria medication), has been greeted with high hopes—always to be shown, with time, not to be as effective as first hoped, and often to cause serious side-effects. But because the consequences of arthritis are so grave for so many of its victims, the search for new drugs continues, and patients and doctors continue to hope for the discovery of a wonder drug.

Some long-term arthritis victims are so desperate for relief that they apply to the skin surrounding their stiff joints DMSO (dimethyl sulfoxide), an oily chemical marketed as an industrial solvent. The Food and Drug Administration has approved DMSO for only one medical use, to relieve the pain of a rare bladder disorder. The Arthritis Foundation has stated that proof of DMSO's effectiveness is lacking, and has warned against the unsupervised use of some DMSO preparations. But in the late 1970s, with no federally approved wonder drug on the horizon, DMSO became extremely popular for thousands of chronic arthritis victims.

Then, on May 19, 1982, the Eli Lilly Company of Indianapolis, Indiana, which had sold $1.6 billion of pharmaceuticals the previous year, introduced a new antiarthritis drug in the United States. It was extolled in ads and in the medical literature as the long hoped-for answer to the disease. Doctors reading both the ads and the medical literature were understandably impressed.

Oraflex (benoxaprofen), as the drug was called, is basically an anti-inflammatory drug unrelated to cortisone—one of the most potent and widely used anti-inflammatory agents in the entire field of medicine. Cortisone can have dangerous side-effects. Oraflex was described as a relatively safe and dramatically effective drug for arthritis. It was just what everyone was awaiting, and doctors rushed to prescribe it and patients to take it—until some who took it died.

The FDA had given the Eli Lilly Company approval to market the drug a year earlier, in April 1981. In evaluating the effectiveness and safety of Oraflex, the FDA followed its standard procedures. Eli Lilly submitted data gathered from animal experiments that showed how the drug worked, and information from four thousand volunteer patients that purportedly showed that it was effective and essentially safe.

As with all applications it receives for permission to market drugs, the FDA left the burden of proof to the manufacturer. Thus, the FDA's decision to let Lilly market Oraflex was based on a scientific review of the information presented by the company, which had a $50 to $75 million stake in successfully bringing Oraflex to market (the amount of money the company spent in developing the drug). Many have criticized this method, charging that the FDA should reach out and seek additional information, but FDA officials argue that they must operate strictly according to the Food, Drug and Cosmetic Act, passed in 1958. Moreover, the FDA has not been given the resources for independent investigations of every drug submitted for approval.

At the time Oraflex was introduced, Lilly stated that it could cause liver damage, but only rarely. But in Great Britain, where the Lilly product had been on the market for two years under the name Opren, doctors were finding that this side-reaction wasn't so rare at all. Within a month after Oraflex was released in the United States, it became known that it had caused twelve documented deaths in Great Britain. The slowness of the British Health Ministry's system for detecting adverse reactions to licensed drugs came under heavy fire from some members of Parliament, who argued that the Committee on the Safety of Medicines—the British equivalent of the FDA—should have noted the association between Opren and fatal side-effects long before.

Alert to the twelve deaths reported from Britain in June 1982, Dr. Sidney M. Wolfe, director of the Nader-founded Health Research Group (and author of the best-seller *Pills That Don't Work*), petitioned Richard S. Schweiker, Secretary of Health and Human Services, to ban Oraflex. Schweiker did not act on Wolfe's request (nor did British authorities do anything about Opren, citing a lack of sufficient evidence of its danger). On July 22, 1982, Wolfe again petitioned Schweiker, this time citing the death of a forty-seven-year-old Nevada woman who had taken the drug for only twenty-nine days. Wolfe presented British statistics, now revised to include a total of forty-five deaths in that country over a twenty-month period. He again asked Schweiker to ban Oraflex as an imminent hazard to public health.

Formally petitioning the FDA, the Health Research Group, joined by the fifty-thousand-member American Public Health Association, cited the British experience in detail. Nineteen Oraflex users had died from either perforated ulcers or gastrointestinal hemorrhage, and nineteen more from liver or kidney damage. The physician of the Nevada woman who had died, Dr. Michael Rask, said that while the drug had helped her arthritis it had killed her. An autopsy found that her liver had been severely damaged. The patient had not been on any other drugs and had otherwise been healthy—the link between her damaged liver and Oraflex was indeed strong.

The urgent calls for a ban on Oraflex from the Health Research Group and the American Public Health Association met with a lethargic, bureaucratic response. Lilly reassured consumers, and the FDA claimed that it was studying all the data. Meanwhile, people were getting sick and some were dying. The FDA cautioned Oraflex users not to draw conclusions from the sort of raw data cited by the Health Research Group.

Dr. Edward Nida, an FDA spokesman, said that "If a patient is concerned, he should consult his physician about the other twenty drugs in the same class." An Eli Lilly spokesperson said that the drug was no riskier than aspirin when properly prescribed. And Dr. Ian Shedden, vice-president of Lilly's research laboratories, said that "All drugs in the benoxaprofen class are associated with deaths from time to time. The considerable benefits of the drug must be assessed with the risks."

What he didn't mention, of course, were the considerable financial benefits to Eli Lilly of successfully marketing this "answer" to arthritis. If Oraflex succeeded, Lilly stood to reap a financial windfall. Annual sales of Oraflex could reach $250 million by 1985, predicted David H. MacCallum of the stock market firm of Paine Webber. What Librium and Valium had done for Hoffmann-LaRoche—another giant pharmaceutical company—Oraflex could do for Lilly. Or so, at least, many Lilly officials hoped.

Amidst all this furor, Lilly did take one small, concessionary step. On June 29, 1982, it sent a circular letter to physicians, pharmacists, and hospitals reemphasizing its recommendation that elderly patients with impaired liver and kidney function be given only one-half to two-thirds the normal dose of Oraflex.

The general public was understandably confused. A *New York Times* headline of July 25, 1982, succinctly stated: "Arthritis Drug Ban: Pro, Con and Undecided." In an interview carried on the same page, Dr. Arthur Hull Hayes, Jr., the FDA commissioner, clearly conveyed the notion that as far as drugs were concerned, some risk must always be accepted.

This position was disturbing to many health and consumer groups, but it was not surprising, given the commitment of the Reagan Administration to deregulation and weakening of the powers of some regulatory agencies. In fact, just at the time that the Oraflex story was making headlines, the FDA was moving ahead with plans to "streamline" the approval of new drugs. The FDA had proposed changes that would reduce the amount of paperwork submitted with new drug applications by as much as 70 percent, and in many cases bring the drugs to market much sooner.

The reason for this, the agency argued, was that it takes American drug companies from seven to ten years from the time they begin testing drugs on humans until the drugs receive final approval for marketing. During this time the companies must submit 100,000 to 250,000 pages of documents, which the FDA often takes two or more years to review. The result is a so-called "drug lag," in which potentially useful new drugs are brought to market in this country long after they are marketed abroad. Critics feel that many Americans suffer or even die awaiting the completion of tests and paperwork on these drugs.

The pharmaceutical industry had been advocating an FDA speed-up for years. Sidney Wolfe would later charge that the Oraflex case was the first example of what would happen with such an accelerated review.

Although the FDA's long-established procedures are ponderous, they are by no means efficient. During their review of Lilly's Oraflex application, for example, FDA investigators had overlooked reports on jaundice that had been submitted as an addendum to the original application. The manufacturer had provided the information, but the FDA had misplaced it. The disclosure of this oversight prompted members of the Congressional Subcommittee on Intergovernmental Relations to wonder if the FDA had overlooked more severe side-effects of Oraflex. But Commissioner Hayes testified that had his agency known of the reports of jaundice, it would still have approved Oraflex; it would have merely required that jaundice be listed as a side-effect of the drug on labels and in information provided to physicians.

Despite Hayes' attempt to put the best possible face on a grim situation, members of the subcommittee as well as many others expressed serious concerns about the FDA's competency, and how drug safety might be eroded by the proposed streamlined review process, which would replace detailed case reports with summary data, and aim for drug approval in six months instead of two years.

Another problem exemplified by the Oraflex experience is that of how drug companies promote prescription drugs. The Eli Lilly Company had launched an enormous advertising campaign aimed at the 31 million American arthritis victims in the hope of capturing a large part of the $1 billion-a-year arthritis-drug market. In this campaign, Lilly made false and misleading statements about the safety and benefits of Oraflex. The FDA criticized these portions of Lilly's campaign and demanded that they be corrected. Among Lilly's scientifically unsubstantiated claims was that Oraflex was notably different ("a new direction," the company called it) and better than the dozen or so similar drugs in the same chemical class already on the market. Lilly also downplayed the seriousness and character of some side-effects that appeared in patients using Oraflex who were exposed to sunlight. The company characterized two of these side-effects—onycholysis, a loosening of the nails from the nail beds, and a severe rash—as "mild, transient and avoidable." In a letter to Lilly, the FDA pointed out that two-thirds of the people taking Oraflex developed a rash, and that neither of the side-effects was mild, transient, or avoidable. The FDA ordered Lilly to send a second news release to everyone who had received the inaccurate one.

By this time, however, Lilly's promotional efforts among the general public had led to an enormous demand for Oraflex from arthritis victims. Physicians were beseeched by patients, and during the twelve weeks Oraflex was on the market, nearly 500,000 persons received prescriptions for the drug. Lilly's consumer-oriented promotional campaign angered many physicians, who reported that patients exposed to Lilly's advertising campaign were pressuring them into prescribing Oraflex, and that without such pressure they wouldn't have recommended the drug. Lilly's hard sell to consumers led to a physicians' "backlash": more reports of Oraflex side-effects (according to the FDA) than might otherwise have occurred.

Although the FDA has long monitored prescription drug advertising aimed at physicians, it has only recently moved toward regulating pharmaceutical companies' press releases, which—along with interviews with the companies' scientists—are a powerful tool for promoting sales. As any practicing doctor knows, patients are increasingly asking physicians to prescribe drugs they have learned about from the newspapers or television, and this is especially true among patients suffering from such chronic conditions as arthritis.

On August 4, 1982, the British Health Ministry announced that it was suspending all sales of Opren for ninety days, pending a thorough review of the evidence gathered by the Committee on the Safety of Medicines. At this point, the committee knew of sixty-one deaths among users of the drug, along with 3,500 reports of adverse side-effects.

This belated action of the British authorities forced Secretary Schweiker's hand. Later that day, he announced that Lilly had "agreed to voluntarily suspend its sales and distribution of the anti-arthritic drug Oraflex." By this time, it appeared that nine people in the United States had died from using the drug. Officials of the Eli Lilly Company had met earlier in the afternoon with the FDA chief to work out the voluntary withdrawal. News of the British suspension would have been banner headlines the following day, putting Schweiker, Hayes, or both in untenable positions if they didn't work fast. FDA spokesman William Grigg told *The New York Times:* "Maybe they decided they didn't want to get hit over the head with tomorrow's headlines."

In withdrawing Oraflex, the Eli Lilly Company offered no *mea culpas*. It stood behind Oraflex as safe and effective, noting that it had been tested with four thousand volunteers and observed for the two years it was marketed in Germany, South Africa, Great Britain, and Spain. But on the same day, Sidney Wolfe revealed devastating information which, if confirmed, would show that Eli Lilly had knowingly imposed a serious risk on American arthritis victims.

Wolfe said that Eli Lilly had known about twenty-six British deaths *before* the FDA approved Oraflex for use in the United States. Lilly hotly denied this, saying that it did not have the resources to solicit this kind of information in the countries where the drug was being used. (The company portrayed itself as the victim of what it called an "environment of hysteria.") But as time and investigation would show, there was considerably more to Wolfe's allegation than Eli Lilly cared to admit.

Representative L. H. Fountain of North Carolina held hearings on the FDA's approval of Oraflex, and on September 20, 1982, announced that Eli Lilly's British subsidiary knew of eight deaths from the drug and had, in fact, reported them to British authorities months before the FDA approved Oraflex for sale in the United States. Lilly denied that it knew of any deaths before the drug had been marketed in the United States.

Several weeks later FDA investigators inspected Lilly's records in Indianapolis, Indiana, the company's headquarters. The FDA's report of this investigation, according to *The New York Times*, stated, among other things, that Lilly "did not promptly investigate and report to the FDA all findings associated with the use of the drug." The report went on to say that twenty-six British deaths "were known to Eli Lilly and Company and/or its divisions, subsidiaries or affiliates prior to the new-drug approval date." The report also charged that Lilly failed to report these deaths to the FDA even *after* Oraflex had been approved for use in the United States.

Responding to these grave charges, E. Ronald Culp, a Lilly spokesman, said, "The company has complied with the laws regarding the reporting of adverse reactions." Other Lilly officials tried to get the company off the hook by saying that they didn't know that American pharmaceutical companies *had* to report to the FDA information received from overseas affiliates. Agency regulations, however, do require such reports. And when Dr. Ian Shedden of Lilly Research Laboratories said in court depositions that the deaths were not reported to the FDA because they were to be expected with that type of drug, Dr. Paul H. Plotz, a senior medical investigator for the National Institute of Health, responded: "The concept of medically inconsequential deaths is something foreign to me."

Eli Lilly's heralded arthritis drug venture eventually took forty-three lives in the United States and ninety-six overseas. The FDA referred the matter to the Justice Department for possible prosecution. By mid-1983 the Lilly Company was the subject of over eighty lawsuits launched by people who had used Oraflex. In the first case to come to trial, a federal jury in Columbus, Georgia, awarded $6 million to a man who charged that his mother had died after taking Oraflex. The company's legal position was not bolstered when Lilly reported in December 1983 that a two-year study involving Oraflex showed an increased incidence in liver cancer among laboratory mice.

There are many reasons why a drug company would impose a risk on the public. The need to reap income to defray costly research and development is certainly at the top of the list, but honest errors in assessing the relative risks of the drug and doubts about the reliability of case reports of its effects from abroad are among other possible explanations.

Nor do company scientists—aware of the multimillion-dollar costs being borne by their employers—find it easy to determine the risks posed by a possible new drug. One can understand why scientists in such a position might give a drug the benefit of the doubt. Defects in the design of the human volunteer studies of the drug further complicate the risk-detection process, and the studies may also not be large enough to reveal the true levels of risk.

Whatever the reasons, in the case of Oraflex, a major drug company did succeed in imposing a significant risk on the American public, despite the existence of the FDA.

The Oraflex case did not set back the FDA's call for a speed-up in the processing of new drug applications. Proposed changes for implementing this speed-up were published in the *Federal Register* in October 1982 and June 1983, and the prospects were good that they would soon go into effect.

Sidney Wolfe was outraged. He said that the Oraflex case demonstrated that the process "should be made more thorough instead of less so." Wolfe explained that the drug-lag is "generally a benefit to this country because it gives us information on the safety of drugs being marketed abroad before we take the risk at home." He argued that a better way to speed the drug-review process would be to increase the size of the FDA's investigative and clerical staff, rather than watering down the process itself.

Another FDA critic, Dr. Lonnie B. Hanauer, of the Arthritis Foundation in New Jersey, noted that if the drug lag had kept Oraflex off the market for an additional three months it would never have been released. "At least eleven patients in this country would not have died," Dr. Hanauer said, "and no patient would have missed out on any important or enduring relief."

Yet the Oraflex experience is not the most dramatic example of the possible effects of a "drug-rush." In one famous case, a new drug, introduced in Germany in 1956, quickly became the most popular sleeping pill on the European market. It was considered so safe that it was made available without prescription, and was specifically recommended for pregnant women. But when its American distributor put pressure on the FDA to approve it quickly, one suspicious FDA physician, Dr. Frances Kelsey, refused to rush to judgment, claiming that the company had not provided enough information on the drug. By 1961, more than eight thousand deformed children had been born in Europe to mothers who had taken Thalidomide. Americans had been spared.

The lesson of Thalidomide, however, seems to be fading. "It is morally wrong," Commissioner Hayes said following the Oraflex affair, "for effective drugs not to be taken to market as fast as possible."

This philosophy, in which human beings often serve as guinea pigs, would be more viable if industry and the government took steps to set up a system for monitoring the performance of new drugs, so that both FDA and industry mistakes could be caught quickly. As the Oraflex example shows, many people can die before reports of the adverse effects of a drug become sufficient to prompt official action. A 1982 FDA survey revealed that three-quarters of patients get no information from their physicians on possible side-effects of the drugs they take, and many patients—and even some doctors—never even consider linking a medical problem with a drug that has been prescribed, believing that if a drug is approved by the government it must be safe.

No drug is totally safe. Yet since 1980, the recommendations of a commission, created by Congress, calling for a new system of monitoring marketed drugs, have gone unheeded.

Close-Up (2)
IRRADIATION

"You eat, you die. You don't eat, you die. I'd rather eat."
—Eric Bogosian, comedian

The average American eats 1,500 pounds of food a year—1,491 pounds of *food* and about nine pounds of chemical additives. These include pesticide residues, preservatives, flavoring agents, coloring agents, and stabilizers.

Several thousand different chemicals fill these roles. Some toil in anonymity, as a result of packaging regulations that do not require them to receive billing; others become quite famous. In the past several years, cyclamates, saccharin, Red Dye #2, nitrates, and nitrites—to name just a few of the famous ones—have been exposed, studied, and in some cases, banned.

But few additives receive star treatment. In most cases, consumers are not even aware of their presence; in others, their presence is well known but well accepted, either because it has become traditional (after many years) or because it is almost unthinkable to exclude certain products from our diet.

In this day of food packaging emblazoned with the word "Natural" to dispel doubts about wholesomeness, some food products seem so natural to begin with that we never think about what may be lurking beneath the surface—or, for that matter, in plain view. Take the orange, for example.

Just to get oranges to the picking stage, growers barrage their groves with herbicides and pesticides, some poisonous (yet within government standards). Many consumers accept the herbicide/pesticide cycle as necessary for creating a bumper orange crop. Few are aware, however, of what happens to an orange after it is harvested. First it is sprayed with fungicide to retard rotting. Then, to meet consumer expectations, the naturally greenish Florida orange is dyed orange by using ethylene gas and the citrus dye known as Red #2, which some studies have linked to bladder cancer in laboratory animals. After this, a petroleum-based shellac or wax is applied to its skin.

This amounts to a triple-threat for orange-lovers: fungicides, dyes, and waxes. The fungicides, it is said, are used in very low quantities and do not penetrate the skin of the orange. The FDA permits Red #2 to be used on oranges only in small amounts—two parts per million by weight—but consumer groups charge that even this level is too high because of its cancer-causing effects in animals; however, growers dispute this, and point out that few consumers eat orange peels.

The waxes on dozens of kinds of fruits and vegetables—from apples to zucchini—are similar to those used to polish automobiles and shine floors. They often contain petroleum, paraffin, and shellac, as well as polyethylene, synthetic resins, or both, and they cannot easily be washed off because of their high melting temperature. The FDA contends that the waxes are harmless when eaten because the molecules are too big to be absorbed in the stomach, but this premise has not been widely studied, since waxes are among the many substances on the FDA's "generally recognized as safe" (GRAS) list, which do not usually undergo rigorous testing before being used for a new purpose. Some such "safe" items, however, as we have noted, have fallen off the list. The safety of ingesting shellac is still under review. Since it is used to coat jelly beans, "the President [Reagan] is our number one guinea pig," Corbin I. Miles, head of the FDA's review branch, said in 1982.

Why does the poor orange—so tasty in its natural state—have to go through so much (largely cosmetic) dyeing, coating, and buffing before it can reach the consumer's hand? The coloring is needed, Florida growers point out, to compete with naturally orange-colored oranges from California and Israel, and cite the test-marketing of naturally green Florida oranges in Canada twenty years ago—which was an abject failure—as supporting their claim. As for the waxing, while it does have some practical use in retarding the evaporation of water and vitamins, it is usually added simply "to increase sales appeal," according to the United Fresh Fruit and Vegetable Association.

But even if the risk in eating oranges is not great, few consumers are aware that there is so much as a sliver of danger, for while growers generally comply with labeling laws by listing the waxes, chemicals, and colors used on their products, retailers generally fail to display the cases bearing the labels or to post appropriate notices in their stores. Maurice Guerrette, a New York State food inspection official, has said that his department is not even checking for labeling violations, instead using its limited resources to inspect the food itself. And a recent FDA bulletin stated that while "we are aware of the problem, [we] don't have sufficient resources to police retail establishments."

Since their use is purely cosmetic, artificial food dyes are the most regrettable commercial additives in view of their apparent ability to cause cancer or damage the internal organs of laboratory animals. (More than a dozen food colors have been banned in the past fifty years.) In fact, most of the synthetic food dyes used today are suspected of causing health problems, including the most popular of them all, Red #40, which replaced a series of banished predecessors and is now itself accused of promoting the growth of cancers in mice. Yet more than 10 percent of the foods we eat in America are artificially colored.

But some potential food dangers pose a far greater risk than coloring agents.

In 1979 the EPA banned the herbicides 2,4,5-T and silvex, which in most cases are contaminated by dioxin. However, even though the agency said the herbicides posed "risks which are greater than the social, economic and environmental benefits" stemming from their use, it made an exception with regard to rice paddies and rangelands. And so both herbicides are widely used in rice paddies such as the ones near Stuttgart, Arkansas— the self-proclaimed rice capital of the world. Arkansas rice is sold throughout the United States and abroad, but it has never been analyzed for dioxin, either in the field or on the way to the market. Yet fears about the effects of the herbicides on Stuttgart's local population have risen in recent years as statistics have shown that the cancer rate in the area is about 50 percent higher than the national average.

For years scientists and writers have warned that the pervasive practice of adding antibiotics to the feed of healthy cattle (to promote growth) could have astonishing repercussions. In theory, some germs in the animals would eventually grow resistant to antibiotics, and when beef from the animals was fed to humans, it would cause diseases that could not be treated with antibiotics. This remained theoretical until September 1984, when doctors at the Centers for Disease Control traced an outbreak of food poisoning in the Midwest, which took one life, to drug-resistant germs. The doctors declared that their study "emphasizes the need for more prudent use" of antibiotics in animals. Although many European countries have virtually banned antibiotics as feed additives, the Reagan Administration (under strong lobbying pressure from drug companies) attempted to relax standards in this country. About half of the antibiotics produced in the United States are fed to animals, and there is some evidence already that the effectiveness of penicillin and the tetracyclines in the treatment of human illness has declined.

Unsettling examples like this could fill this book. Did you know that newly harvested potatoes that must be stored for long periods are often chemically treated, and that one of the two chief chemicals used for this purpose is maleic hydrazide, a suspected carcinogen? Almost every week produces a new scare story involving a previously anonymous pesticide, preservative, or artificial food coloring. In April 1983, for instance, a new Washington, D.C., consumer group called Public Voice charged that the FDA's pesticide detection program is "virtually useless as a tool to prevent the sale of excessively contaminated produce to consumers." And no wonder: there are at least thirty-five thousand pesticides on the market, containing about 1,400 active ingredients. Former FDA Commissioner Donald Kennedy once observed: "By the time we have completed our analysis of a product and found an illegal residue . . . the product has already been distributed in the marketplace."

Consumers have grown increasingly concerned about these practices. A Harris Poll survey taken in January 1984 for the Food Marketing Institute revealed that 77 percent of those polled considered chemical residues in their food as a serious hazard; only 45 percent felt the same way about cholesterol, 37 percent concerning salt, and 32 percent in regard to preservatives. The survey also revealed that the number of consumers who relied primarily on themselves to be sure that a product was safe was 48 percent.

However, a controversial process that could mitigate many food hazards—both hidden and blatant—may be right around the corner. Deemed safe by the FDA, and then widely used by the food industry, it could reduce the post-harvest application of insecticides that are known to be carcinogenic. It would eliminate the need to cure meat with nitrites and nitrates— substances that can form carcinogenic nitrosamines when heated. It could cut spoilage, which now ruins an estimated 25 to 30 percent of the world's food supply. And it would extend the shelf life of products for weeks, increasing the length of time food could be stored (or transported) and still taste "fresh," while decreasing the need for many of the chemical preservatives now used for this purpose.

This may sound too good to be true, but it is not; all of these wonderful advances could be accomplished practically overnight. The only hitch is this: The revolutionary process that would produce these changes may be hazardous itself.

The process, irradiation, involves conveying food in any form— from a bunch of bananas to a bag of wheat—through a shielded chamber, where it is briefly exposed to gamma rays emitted from a radioactive substance such as cobalt 60 or cesium 137. Gamma rays, which are shorter and more penetrating than X rays or microwaves, kill Salmonella and other bacteria by disturbing their metabolism, making them unable to divide and grow; the rays also curb the ripening of vegetables and fruits by slowing down cell division.

The food itself does not become radioactive—just as teeth do not become radioactive from dental X rays. Onions can stay fresh for a year after exposure to gamma rays; chicken and fish can be refrigerated for three weeks or more and still taste good.

Irradiation is desirable not only because its effects are so long-lasting, but because it "cold" processes food. This means that the gamma-ray bombardment raises the temperature of the food only slightly, which, in turn, keeps such adverse changes as altered color, odor, flavor, texture, and loss of nutrients at a minimum. (Most food that is heated and then canned must first be packed in liquid, and becomes "squishy"; food that is frozen suffers a textural change.)

But irradiation is more than a technique to preserve foods; it is also a way of achieving a variety of effects in the finished product. For example, beans can be irradiated to reduce their cooking time; meat can be tenderized; irradiated wheat provides a bigger loaf of bread than nonirradiated wheat. Someday, irradiation may even have a major impact on the imbibers of wine and beer. Research in Hungary indicates that irradiated grapes yield up to 20 percent more juice than nonirradiated ones, and brewing industry research reveals that irradiated barley increases its yield by seven percent during the malting stage.

Opponents of irradiation, however, charge that too little is now known about the process to give it a green light, and FDA and food industry representatives continue to encounter consumer skepticism when they talk about irradiation in public. They complain that the only thing people seem to want to know about irradiated food is: "Will eating it make me glow in the dark?"

Nevertheless, in February 1984, the FDA, as a first step toward the wide commercial use of irradiation, proposed a regulation to allow its limited use for killing bacteria on spices and insects on fresh fruits, vegetables, and other foods, and for retarding spoilage. This proposed rule does not cover meat, poultry, or fish, and has no effect whatsoever on the broader uses of irradiation as a preservative.

"It merely starts the regulatory process," says Jim Greene, an FDA spokesman. "The speed at which we go now depends on the food industry. Whether it flies or not is up to the industry."

Irradiation is nothing new. Franklin S. Smith of Philadelphia obtained a United States patent in 1909 to use X rays against a tobacco pest. A German named Otto Wust filed for a French patent after discovering, in 1920, that ionizing radiation could preserve food. In 1943 the U.S. Army asked the Massachusetts Institute of Technology whether food for our fighting men could be safely preserved by irradiation, and four years later M.I.T. scientists said "yes." In 1953, during the Atoms For Peace period, the Army started a wide range of irradiation experiments at its laboratory in Natick, Massachusetts. Over the next twenty-five years the Army spent $50 million on studies which indicated, officials said, that food irradiated at low levels tasted fine and was safe to eat.

Nobel Prize–winning chemist Willard Libby told his colleagues on the Atomic Energy Commission that food irradiation was the most important technological development in food processing since canning came into commercial use in the 1880s. "The millions who live in the shadow of starvation deserve the benefits of more rapid introduction of irradiated foods," Libby said. "Wasteful as man has been in the past, he can no longer afford to waste time, technology and food."

The AEC judged that the irradiation process "has been more thoroughly tested than any other method of food preservation."

The results, however, were not overwhelmingly positive. Meat turned mushy during the process and, according to one taster, tasted like "wet dog," and researchers kept coming across mysterious substances in irradiated foods which they eventually called "URPs"—unique radiolytic products. These URPs are simply microscopic substances that would not be present in foods that had not been irradiated. This does not mean that these substances are harmful, but researchers have still not been able to identify all of them or say for certain that they cause no ill-effects. (A recent internal FDA memo puts it this way: "The safety of radiolytic products has been a continuing problem to FDA.")

And so with energy cheap, food in America plentiful, and frozen food in vogue, irradiation was abandoned. In 1958, Congress redefined irradiation as a "food additive" rather than a food-processing technique, which meant that the process had to face much more stringent control and study. The fledgling irradiation industry died.

Overseas, however, irradiation marched on. To date, more than twenty countries have approved the use of forty kinds of irradiated foods, experimentally or for human consumption. In most cases, the only foods covered are onions, potatoes, or garlic, although irradiated shrimp and strawberries are being test-marketed in the Netherlands, and irradiated poultry and fish fillets in Canada. A 1981 report by the World Health Organization said that there was no evidence of adverse effects from food subjected to low doses of radiation, and that further toxicological testing of such foods was unnecessary.

In the United States, the irradiation of wheat (to kill insects) and potatoes (to inhibit sprouting) has been legal since the early

1960s, but has not been widely utilized because these crops are so abundant. Irradiation has been used experimentally, however. During the Apollo space program, men on the moon ate ham sterilized by irradiation, and Soviet cosmonauts taking part in the 1975 joint Apollo-Soyuz mission complimented their American counterparts on the tasty (irradiated) beefsteak they had brought along. Irradiated food has also been approved for hospital patients who must eat sterilized food. Dr. Sandra Aker, director of clinical nutrition at the Fred Hutchinson Cancer Research Center in Seattle, Washington, has been irradiating foods for cancer patients since 1974. In other areas, the FDA granted a California company an emergency permit for an irradiation plant to combat the Mediterranean fruit fly infestation, and has okayed the exporting of irradiated food to countries which permit its sale. And irradiation is also widely used on nonfood products—from baby powder to Teflon on cooking utensils.

But it was not until 1981 that the FDA reconsidered the popular use of irradiation. On March 27, 1981, it announced that it was "considering a new policy to facilitate the use of radiation to preserve foods." The agency suggested (for the time being) a "safe," 100-kilorad ceiling for irradiation dosage, well below the World Health Organization's proposed 1,000 kilorad standard. (One kilorad is a dose roughly one hundred thousand times greater than a person would receive from a normal chest X ray.) To get the full benefit of irradiation as a preservative, however, the dosage will have to climb much higher. How high it climbs will be the number to keep an eye on as the expected irradiation debate develops.

At doses under 100 kilorads—the current FDA limit—irradiation can reduce sprouting of vegetables such as potatoes and slow the ripening of fruits. At over 100 kilorads, the effects are more interesting. With an 175-kilorad dose of gamma rays you can slow down the growth of mold on berries. Crabmeat bombarded with radiation at 350 kilorads will stay fresh for about forty days. At 500 kilorads you can eradicate Salmonella bacteria in frozen shrimp, and at 500 to 700 kilorads you can get rid of these bacteria in poultry. All of these uses fall below the international 1,000 kilorad (one megarad) standard safe "ceiling" dose for irradiation—itself a giant step below the 4.3 megarad dose that completely sterilizes chicken and makes it last indefinitely without refrigeration.

Nevertheless, the FDA's 100-kilorad proposal would get irradiation off and running in America. One factor which may have tipped the FDA toward reconsidering irradiation was the finding, by EPA scientists in 1980, that a leading grain and citrus fruit fumigant, ethylene dibromide (EDB), was a "known carcinogen and powerful mutagen." Maureen Hinkle, a scientist with the National Audubon Society, said that tests on animals suggested that the substance posed "the highest risk of [causing] cancer of any chemical we know of." Environmentalists charged that the EPA had not acted to bar the use of EDB because the Reagan White House had intervened on behalf of grain and fruit growers and chemical manufacturers to keep it on the market. (It was finally banned for agricultural use in September 1983.)

A number of scientists have emphasized that irradiation is the only known, effective replacement for EDB. Niel Nielson, president of Emergent Technologies, Inc., in San Jose, California, charged that any delay in coming up with a new irradiation policy put a heavy responsibility on the FDA "for each and every additional cancer problem created by the use of EDB."

Following the FDA's 1981 proposal for the use of irradiation at doses up to 100 kilorads—known as an "advance notice"—the agency received more than 100 letters from scientists, food manufacturers, and private citizens, among others. The range of views was predictable. A commentator from Nebraska said that safety problems surrounding the use of irradiation "appear to be nil," while someone from Van Nuys, California, said, "I don't care how low-level the radiation is, it contributes to cancer."

The advance notice (followed three years later by the FDA's proposed rule for allowing the irradiation of spices, fruits, vegetables, and some other foods) created "a lot of interest in the public and scare stories in the media," according to the FDA's Jim Greene. But, he says, it also sparked considerable interest in food-industry circles. The industry began to envision strawberries on fruit stands remaining bright red and firm for weeks; fresh fish being shipped deep into rural areas; and food preserved without the need for freezing, canning, cooking, drying, smoking, or using chemicals. A seminar on irradiation at Rutgers University in May 1982 attracted representatives from several major food companies, including General Foods, Nabisco, Campbell Soup, M&M/Mars, and Allen Products, which makes Alpo dog food. (Irradiating dog food has not been permitted by the FDA because poor people sometimes eat dog food.)

"Any time a new process becomes available it's exciting for the food industry," says Dr. Myron Solberg, a food scientist at Rutgers and an irradiation advocate. "The renewed interest they are showing is clearly profit-oriented. Let's not kid ourselves. They don't want the other people in the industry to get ahead of them."

While some observers feel that irradiation may be prohibitively expensive for many companies, others disagree. "We wouldn't be in this darn thing if we didn't think it was profitable," said George Giddings, who works for Radiation Technology, a Rockaway, New Jersey, company that irradiates food for American astronauts and for export. According to Giddings, the cost of processing fish and chicken to make it last an extra week or two on the shelf is only an extra one or two cents per pound. And irradiating consumes much less energy than heating and fumigating.

But Dr. Martin Welt, who founded Radiation Technology, wonders whether the food industry will ever strongly pressure the FDA to make irradiation commonplace. He sees little reason for the industry to do this as long as it is permitted to use chemical fumigants and nitrites, and as long as the presence of Salmonella and other disease-causing microbes in food is tolerated. "They're selling all the food they want," Welt says, "and if it happens to contain Salmonella or carcinogens, they sell it anyhow. The government doesn't stop them."

Thus, the irradiation picture is not altogether rosy, and is further clouded by the fact that the process does have limitations. Doses high enough to completely sterilize food and create meat and poultry that require no refrigeration do alter the flavor of these foods. Fresh milk can't be irradiated, since it sours; and some fruits and vegetables subjected to irradiation end up with a badly bruised look. And questions about the safety of irradiation remain.

Sanford A. Miller, director of the FDA's Bureau of Foods,

said in 1982 that irradiation "causes no overt changes in food, only subtle changes"—the creation of those new, "undeciphered" chemical compounds known as URPs. But in April 1982 an FDA staff scientist (who asked that his name be withheld) told *The New York Times* that irradiation research was inadequate, and that the long-term health consequences of the process had not been examined. He contended that studies of irradiation had been flawed, inconclusive, or biased in favor of the new technology. And he charged that even "subtle" changes caused by the irradiation of foods could be hazardous. More study, said the scientist, was needed to determine the nature, quantity, and effects of URPs. "We've got to think of the effect of these new chemicals on the lives of generations and generations to come," he said.

National consumer groups have not yet taken a strong stand on the irradiated food issue. Some are less concerned with irradiated food than with the big boom it might create in the radiation industry. The widespread use of irradiation, they argue, would put at risk thousands of workers who would handle the irradiating equipment; threaten homeowners who live near food factories; lead to a great upsurge in the volume of radioactive material being transported over highways and through towns and cities; and further add to this country's already acute radioactive waste problem.

A key factor working in favor of irradiation, however, is that the processes and products it would replace—such as EDB—are themselves harmful. This helps balance irradiation's "cost-benefit ratio." Doing away with nitrites and nitrates and with toxic or carcinogenic preservatives and chemical fumigants would be a healthy step forward. And, of course, keeping food from spoiling, in a world in which enormous numbers of people go to bed hungry, is no small thing.

However, when consumers face the choice of whether *they* will eat irradiated berries or two-week-old "fresh" irradiated trout, the choice will become more personal, while global hunger and other concerns fall away. In an age when more and more people are reading labels closely—and beginning to shun even such dietary staples as sugar and salt—will they purchase a product if it bears the scary new description, THIS FOOD HAS BEEN IRRADIATED? The future of irradiation as a commercial process may rest on the answer to this question.

In a nationwide poll a research company, Consumer Network, found that 98 percent of the two thousand shoppers surveyed said they wanted food to stay fresher longer, but less than half said they would try irradiated meat, poultry, and seafood. But shoppers a century ago would have undoubtedly turned up their noses at the suggestion that they purchase food that had been dyed, gassed, emulsified, chemically preserved, and artificially flavored.

The FDA's Jim Greene says he still feels frustrated at times when trying to explain the merits of irradiation to consumers. "They just don't understand," he says, "that irradiated food does not become radioactive. Of course it won't glow." Edward L. Korwek, a Washington lawyer who specializes in legal issues surrounding irradiation, adds: "Irradiation labeling is an issue because consumers are, to put it bluntly, afraid of the technology."

Some food industry officials feel that FDA regulations requiring the "treated-with-radiation" label would be unfair; after all, they argue, the consumer who now buys canned peas is not told *how* they were processed. One food technologist told the FDA: "The public equates 'irradiation' with 'radioactive,' and therefore the labeling would be misinformation." A seafood processor commented: "The problem we face is the word itself. Consumers do not know how to handle 'radiation.' Finding the word on the retail shelves would cause great alarm in the marketplace."

Some have suggested, as a solution to this image problem, substituting the expression "cold-processed" for "irradiated." But Robert Rodale, the influential chairman of Rodale Press, and a noted critic of irradiation, has pointed out: "If the words radiation or irradiation are ever separated from the process of radioactive treatment of food, you can be sure that the battle to hold that technology at bay has been lost. . . . Once the genie is out of the bottle, no one can predict how large it will grow."

It is worth noting, then, that when the FDA issued its proposals on irradiation in February 1984 its previous labeling requirements were missing. The new rules did not require the labeling of irradiated food products sold in grocery stores. Jim Greene said that the FDA had urged a labeling notice but the Department of Health and Human Services, which oversees the agency, "saw fit not to heed our recommendation on that point."

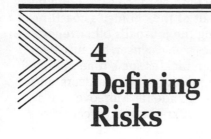

4
Defining
Risks

A risk can be defined as a dangerous factor, as the possibility of injury, or—from the perspective of an insurance underwriter—as the chance of loss. But these descriptions only hint at the real meaning of risk. Risk is, at heart, merely a mathematical measure of the probability and seriousness of harm. It can even be expressed in a formula: Risk = probability × severity.

Mathematics is unpopular with many people, however, and probabilities are frequently viewed as abstract numerical statements that have little meaning for us as individuals. We don't know what to make of the latest disclosures about dioxin, for example, so we insist that scientists and technocrats give us more information. And so researchers agreeably churn out data that are often lost on the public.

This endless search for meaningful information has become so ingrained that it has inspired sarcasm from within its own ranks. Since 1957, scientists who belong to the Society for Basic Irreproducible Results have been publishing a journal (current circulation: forty thousand) that spoofs the modern scientific method. In recent years, the journal has dryly noted—citing skewed statistics—that if everyone keeps stacking *National Geographic* magazines in garages and attics instead of throwing them away, their total weight will soon sink the continent 100 feet, causing extensive ocean flooding. It has also predicted that if beachgoers keep returning home with as much sand clinging to them as they do now, 80 percent of the country's coastline will disappear in ten years. Among the journal's other reports is the startling discovery that pickles cause cancer, because 99.9 percent of cancer victims have eaten pickles some time in their lives. So have 96.8 percent of Communist sympathizers and 99.7 percent of those involved in car accidents. Moreover, those born in 1839 who ate pickles have suffered a 100 percent mortality rate.

While these examples may be outrageous, it is nevertheless true that the public is often unable to distinguish such spurious assertions about risk from solid facts. Yet while people have difficulty comprehending the concept of risk (a negative idea), they deal somewhat more easily with the positive concept of "safety," which actually has to do with the *acceptability* of a risk. But determining this acceptability is a complex matter influenced by many factors, including possible alternatives, the magnitude of the risk, the benefits associated with the risk, and social and personal value judgments.

Something is deemed safe if the risks associated with it are acceptable in the judgment of an individual, a group, or the government. Unfortunately, many people, when they consider something as safe, see it as free from *all* risks (or associated with only minor hazards of rare occurrence). Nothing—whether food, drug, or household appliance—can ever be absolutely free of risks at all times and under all circumstances for all people. Few people, in fact, demand a "zero-risk" society; rather, they want risks to be as low as reasonably achievable (or ALARA, as it's known in the risk-assessment trade).

Obviously, then, there are degrees of risks and also degrees of safety.

Defining risks: Before it can be determined whether or not a risk is acceptable, it must be shown to exist. And more importantly, it must be defined in some quantified manner—and sufficient details about it made known—to enable people to make intelligent choices about it.

There are a variety of methods for determining risk. Some, such as common sense and empirical knowledge, provide "soft information." Not being derived from scientific experimentation, this information may or may not be valid, since it is easily influenced by personal bias and chance. (Bias and chance can also affect improperly designed scientific studies, rendering the statistics derived from these studies no more valid than intuitive opinions.)

In fact, it is common sense rather than scientific research that is ordinarily used first in assessing a risk. It is easy to dismiss such an assessment as invalid, since it cannot be backed up with convincing statistical data. When an empirical or "common sense" approach points to a cause-and-effect process at the root of some risk, the source of the risk (such as a polluting industry) often argues for coincidence. An instance of coincidence might involve someone who took a vitamin pill and then tripped on a sidewalk and fractured his ankle several days later— an event for which the pill was *probably* blameless. Less obviously coincidental is the finding of an empty bottle of sleeping pills next to the bed of someone who has died, quite possibly, in this case, by having committed suicide. When an autopsy shows that the person died of a heart attack, however, and the person's blood does not contain any of the contents of the bottle, it becomes clear that the bottle is a coincidental finding.

Among the factors that often make common sense and empiricism inadequate tools for discerning risks are a long period between the exposure to the risk and the occurrence of damage (such as a disease), which may cloud or prevent any association of the risk with the damage. For other risks, such as smoking and driving while drunk, which are so common that many don't see them as especially dangerous or foolhardy, only scientific or statistical scrutiny can define their magnitude. Many diseases that we may consider common are actually rare when one calculates how many people get them.

Consider, for example, that few Americans know more than 100 people to such a degree that they are familiar with the details of these people's lives. When one also considers that the incidence of lung cancer—which medical scientists define as common—annually affects only about two in every thousand heavy smokers, the chances of knowing someone with lung cancer are seen as being slim. Empirical observations often cannot define—or even recognize—a risk.

Yet even scientific methodologies may have difficulty in defining the relative importance of risk factors in such common diseases as cancer and heart disease, given the number of possible causes of these diseases—from smoking to stress. Nevertheless, risks can be determined with some accuracy, provided the appropriate methodologies are used and pitfalls avoided. Product performance, for example, can be measured during the manufacturing process by testing parts alone and together, after assembly of the manufactured item and before its sale (and even after its marketing). Yet as we know from our experience with automobile recalls, many such testing programs have failed to detect a number of potentially lethal hazards. Often, the failure to detect a hazard prior to the widespread use of a product occurs because the hazard is so rare that it escapes detection in the limited testing to which the product is subjected. Only after large numbers of cars get on the road, for example, do their hazards emerge.

In other cases, the tests themselves are faulty and indirectly lead to accidents and unnecessary illness. This has been the case with many of the new chemicals and drugs that have flooded the market since the 1940s. The federal agencies that have been created to regulate these new hazards commonly set certain guidelines and then demand that chemical and drug manufacturers meet these standards in the products they market. Unfortunately, agencies do not have the resources to test every new drug and chemical themselves, and what is therefore the key link in this safety net—the nation's three hundred independent testing labs, which carry out many of these evaluations—may also be its weakest link.

Only in recent years has it become apparent how weak that link may be. The most dramatic evidence that this is so—which culminated in the trial and conviction (on fraud charges) of three former officials of Industrial Bio-Test (IBT) Laboratories of Northbrook, Illinois, in the summer of 1983—only suggested the dimensions of the problem.

In the IBT case, executives of what was one of the nation's oldest and largest testing facilities were charged with generating false data throughout the 1970s on hundreds of food additives, chemicals, drugs, herbicides, and other substances. From 1952 until it went out of business in 1978, IBT performed some twenty-two thousand studies, including one-quarter to one-third of all toxic tests in the world. Its regular clients included the EPA and FDA, as well as the Dow, Monsanto, and Du Pont corporations, Sandoz, 3-M, and dozens of others. The fraudulent work called into question the safety of nearly 15 percent of all pesticides that had been approved for use in the United States (to mention just one problem area).

The IBT researchers had falsified many test results; others were made out of whole cloth. According to grand-jury testimony, many of IBT's test animals were housed in a waterlogged room known as "the Swamp," where dead rats and mice decomposed so rapidly that their bodies oozed through their wire cages and lay in purple puddles on trays. The death rate among the test animals was so high that it was impossible in some cases to judge the impact of the chemical being tested.

In the wake of IBT revelations the products judged falsely (or not at all) by the lab were not immediately ordered off the market; instead, federal agencies and manufacturers began a painstaking review process. "The public will be at risk during the time it takes to sort out which of these products were wrongly or prematurely judged safe," *The New York Times* noted in an editorial. The EPA contended that it did not have the authority to demand that chemicals which had not been adequately tested be taken off the market. "All we have now are question marks," an EPA official explained. Maureen Hinkle of the National Audubon Society complained that this meant that "the burden of proof still lies with society to prove that the chemicals on the market are not safe."

The IBT scandal was discovered only by accident, when an FDA official, looking for something else, stumbled upon some IBT data that seemed peculiar. How could the companies that were paying IBT to test their products be unaware of the suspicious results—not to mention the shocking conditions at the facility itself? This was not the first testing scandal. Years earlier, the Biometric Testing Corporation was found to have tampered with drug test data. In another case, contractors sent G. D. Searle & Company fudged data on the popular drug Flagyl, used for treating certain parasitic diseases. As early as 1977, Senator Edward Kennedy, following hearings, declared that "millions of Americans are taking products—drugs, food additives and food colorings—which they believe to be proven safe but which, in fact, have not been. It is now clear that our regulatory system has broken down." The Federal Trade Commission in 1978 identified 157 cases of compromised and deceptive product standards and test lab actions, many of which led to injuries and deaths. And in 1983 it was learned that for fifteen months the nation's leading commercial supplier of laboratory animals, the Charles River Breeding Laboratories in Wilmington, Massachusetts, had filled many orders for an inbred strain of mouse with shipments that also contained mice of different genetic backgrounds. Hundreds—perhaps over a thousand—experiments were affected.

Potential conflicts of interest taint test results. Test labs are largely dependent on manufacturers, not the government, for their business; if the labs come up with results displeasing (however accurate they may be) to their clients, they are liable to lose the clients' business; conversely, laboratories that tend to come up with the "right answers" (from industry's viewpoint) often get showered with business. At the IBT trial the prosecutors argued that lab officials began to cover up their errors rather than rerun the tests properly because sponsoring companies were highly competitive and were in a hurry to market their products. The manufacturers may have a less than open-minded approach to this process. "The sponsor of [a] drug," said FDA pathologist Adrian Gross, who uncovered the IBT scandal, "will fight like a cornered tiger to salvage the integrity of the study at issue."

There has been some reform in this field. The FDA has instituted new research standards for drug tests, and has established a team of scientific auditors, now one hundred strong, which "regularly turns up false or suspicious data," according to The New York Times. The Times noted, however, that "since only the most carelessly fabricated data are easy to detect, this may be the tip of the iceberg." In 1983 the EPA, however, still had only one inspector, and had yet to enact laboratory standards four years after proposing them. And as we have seen, many scientists and researchers remain tied to industry in one form or another.

This means, according to many observers, that fundamental reform is needed in the testing process. Some have suggested that private labs be federally approved. Others have called for regulatory agencies, rather than industry, to contract out studies, or have proposed expanding the use of joint industry/government testing facilities, such as the Health Effects Institute in Boston, which conducts highly regarded auto emissions tests. "Nobody in the business of buying laboratory services, if they are smart, goes in for cheap or shoddy work," former EPA Administrator Douglas Costle has said. "But you're always going to have a human element, which means you always need checks and balances." And on this human element rests the safety of the cars we drive, the food we eat, and the household products we use every day.

Defining acceptability: Is this risk worth taking? It can depend on the risk-taker's sensitivity (personal susceptibility), expertise, and experience, among other factors. Consider driving an automobile: it is always somewhat risky, but it is relatively more risky if you have poor vision, if you're just learning to drive, or if it's snowing. (It can even depend on the make of car you're driving.)

It also depends on your point of view. An individual facing an event that is unlikely to occur can be relatively carefree. For example, it has been calculated that the annual risk for each American of dying in an automobile accident is only 2.4 in 10,000— clearly a long-shot chance, and not worth losing much sleep over. But the U.S. government has to look at it another way: these crackups kill almost fifty thousand Americans every year—a clear demand that steps be taken to reduce this threat.

Perspective is only one factor that muddies the waters of risk perception. Mary Douglas and Aaron Wildavsky, in their book *Risk and Culture,* convincingly argue that the perception of risk is actually a societal process because there is no single, correct, *individual* concept of what is and is not risky. They assume that "any form of society produces its own selected view of the natural environment, a view which influences its choice of dangers worth attention." And while "common values lead to common fears," Douglas and Wildavsky note, many dangers are selected for public concern not on the basis of "probability" and "severity" but because social criticism has singled them out. (Douglas and Wildavsky are not very sympathetic to the environmental movement in America.) If asbestos, the authors observe, was created to save people from burning, why is asbestos-poisoning now seen to be "more fearsome than fire"? Why do the French people have so few qualms about nuclear power plant safety, and Americans so many?

Others argue that risk acceptability is usually a political issue. As Philip Handler, former president of the National Academy of Sciences, put it: "The *estimation* of risk is a scientific question—and, therefore, a legitimate activity of scientists in federal agencies, in universities and in the National Research Council. The *acceptability* of a given level of risk, however, is a political question, to be determined in the political arena."

The FDA, for example, has refused to approve the drug Depo-Provera as a contraceptive because it has caused cancers in animals. But it is used in eighty countries abroad, praised by the World Health Organization, and considered by many population-control groups to be the ideal contraceptive (it is highly effective and one injection lasts for three months). Our government continues to feel that Depo-Provera is not an acceptable risk for women in the United States, which has no population control problems, but officials in the Third World feel that its benefits outweigh its risks.

Offering another comment, Philip Handler asks: "Are the estimated risks of nuclear power plants too great to be acceptable . . . are they more or less acceptable than those associated with coal combustion?" A few years ago, after the antinuclear movement had entered the political arena, a physicist at the University of Pittsburgh named Bernard Cohen offered to eat one gram of plutonium if given a chance on national television to explain how safe nuclear energy and nuclear waste management is. Cohen contended that nuclear power plants are no more dangerous to an average person than if that person were overweight and ate one extra banana split during his or her lifetime.

Estimating the causes of cancer is an especially volatile activity and one that often becomes embroiled in politics. Just a few years ago experts were referring to a "cancer epidemic" in America caused by the postwar chemical boom. Ten leading government scientists went so far as to predict that 20 to 38 percent of all cancers in the near future would be caused by just six industrial chemicals. This outlook changed, coincidentally or not, when Jimmy Carter left office and Ronald Reagan arrived. Most influential was a 1981 report by two renowned British epidemiologists, Sir Richard Doll and Richard Peto, who concluded that food additives—and carcinogens in the workplace, in the environment, and in industrial products—account for fewer than 8 percent of the 450,000 American cancer deaths each year. They attributed about 30 percent of cancers to smoking and about 35 percent to diet (excess of fats, lack of fiber, natural carcinogens in food, and so forth). This meant that personal habit—not polluting industry—was the prime culprit.

Many scientists found the Doll-Peto report persuasive, and the new policymakers in Washington (who vowed to raise risk levels, cut regulations, and challenge the public to look out for themselves) naturally utilized it. Those scientists unpersuaded by Doll-Peto claimed that many cancers caused by exposure to chemicals in the past two decades will only begin to turn up in a few years.

The stage was set for Edith Efron and her popular 1984 book, *The Apocalyptics: Politics, Science, and the Big Cancer Lie*—described by some critics as *The Silent Spring* in reverse. Efron denounced scientists who, during the 1970s, substituted "moral concepts for missing data." She blasted public health agencies who, to protect their "turf," misled the public into believing that cancer could be reduced simply by regulating chemicals. And she criticized cancer-fighters who mistakenly believed that lab tests really can distinguish between carcinogens and noncarcinogens. But while Efron, a well-known conservative author, denounced the alleged "political aims" of certain scientists and regulators, her own views reinforced almost point-for-point the cancer policies put forward by the Reagan Administration—policies described as "supply-side carcinogenesis" by some of those whom Efron denounced.

Social and political considerations assume enormous importance because their very existence makes it virtually impossible to devise an objective mathematical method for evaluating the acceptability of risk. One may be able to define risks, but defining their *acceptability* is quite another matter. As Elliot Marshall, a writer for *Science* magazine, has noted, many degrees of prudence and imprudence can be justified— as they were in Caesar's day—according to what one chooses to see in the entrails. "This is why," Marshall writes, "people who try to make reasonable decisions on pesticides and food additives must rely finally on moral and political judgments. They have no choice. Science cannot say how much prudence is enough."

Risk definition has assumed paramount importance in our society because, while the brutal randomness and severity of health risks (such as diphtheria and tuberculosis) have lessened, the absolute number of risks—and public awareness of them—has increased, and the public is being urged to accept many of them. While remarkable technological and scientific advances have settled some old questions about risks, they have raised many new ones. "Our techniques for finding new dangers have run ahead of our ability to discriminate among them," Douglas and Wildavsky write. Fewer than thirty substances are definitely linked to cancer in humans but some 1,500 are carcinogenic in animal tests. Only seven thousand of the more than five million substances known to man have ever been tested for cancer-causing ability. There are some forty-five thousand chemicals now in commercial use, and, according to one estimate, it takes a team of scientists, three hundred mice, two to three years, and over $300,000 to determine whether a *single* chemical causes cancer. Moreover, the necessary use, in defining risks, of mathematical measurements that are poorly understood by many also makes comprehending the risks difficult.

Yet an understanding of how risks are defined is crucial to public decisions about what is and is not acceptable. Often those who enter the controversy over what degree of risk is acceptable—or if indeed any degree of risk is acceptable—do so with a poor grasp of how the risk was defined in the first place. Matters are not helped by the conflict between risk data gathered through laboratory studies, health surveys, and empirical observations. In fact, sophisticated epidemiological tools (such as cohort and case-control studies) can be used to sustain opposing arguments even when applied to the same data base. As a result, differences of opinion about many risks hinge not on facts, but on how they are interpreted.

Risks defined as acceptable are increasingly being urged on the public because certain foods, chemicals, drugs, household products, procedures, and life-styles are perceived as having benefits for most and risks for few. This, coupled with the long period before the consequences of some risk exposures become known and our inability to predict them exactly, works against those who urge strong risk-prevention policies, enabling some unscrupulous manufacturers to continue their risk-producing activities, frequently with little protest from a confused and poorly informed public.

Yet as individuals we can arrive at some comprehension of the risks for a number of possible hazards and, equipped with this information, judge which are acceptable and which are not, and where necessary, take risk-reducing action.

Close-Up (1)
THE PILL
"To put the health risk in perspective, oral contraceptives should be sold in vending machines, and cigarettes should be sold by prescription only."
>   -Dr. Malcolm Potts,
>   International Fertility Research Program

A number of studies reported in the early 1980s that users of birth-control pills were half as likely to develop cancer of the womb and ovaries as those who did not use the Pill. They also showed that the Pill gave considerable protection against salpingitis (pelvic inflammatory disease), which blocks one or both of the Fallopian tubes, through which egg cells travel on their way to the womb. Furthermore, the results of a twenty-five-year study by the Mayo Clinic supported earlier reports that the Pill reduced the incidence of rheumatoid arthritis. In July 1982 Dr. Lawrence K. Altman, *The New York Times'* highly respected medical writer, wrote a major article entitled, "Health Benefits of the Pill Found to Outweigh Its Drawbacks."

Yet only 10 months earlier Dr. Bruce V. Stadel of the National Institute of Child Health and Human Development, in a two-part article in the prestigious *New England Journal of Medicine*, methodically reviewed important studies pointing out some of the dangerous side-effects associated with oral contraceptives, including stroke, heart attacks, high blood pressure, and blood clots in the veins of the leg. And two studies published in the British medical journal *Lancet* showed a strong association between use of the Pill and increased risk of breast and cervical cancer.

What are we to make of this endlessly changing—often conflicting—testimony concerning the safety of oral contraceptives? More than two decades have now elapsed since the introduction of the Pill. For almost that long, medical researchers have attempted to quantify the levels of risks associated with its use. They have had mixed success. It seems, for instance, that the Pill both prevents and causes cancer. Confidence in the safety of the Pill has waxed and waned every few years. This has not been due to new interpretations of old studies, but because fresh research has thrown dramatic new light on the Pill's performance.

For the thirty-six million American women who face the difficult question of avoiding pregnancy, the Pill's widespread use is still a divisive issue. It is undoubtedly true that millions of women have—some with full knowledge, others unwittingly—acted as human guinea pigs in what some critics, such as Dr. Samuel S. Epstein, have called the largest uncontrolled drug experiment of all times. It is also true that the Pill has, at the same time, safely prevented millions of unwanted pregnancies that might have occurred had couples used less reliable forms of birth control.

For the future, the best news is that the Pill's love-it-or-hate-it status may be ending. As more becomes known about how oral contraceptives work—and the estrogen content of the product is adjusted accordingly—it will become easier for physicians, and the public, to judge just who might benefit most from taking the Pill, and who should avoid it at all costs.

The FDA approved the first oral contraceptive, Enovid, marketed by the G. D. Searle Company, in 1961, after only thirty-eight months of tests on just 132 women. Soon thereafter, several other pharmaceutical companies rushed forward with similar products. (This would soon become a $100-million-a-year field.)

The arrival of the Pill was hailed as a major breakthrough in fertility control. The implications for family planning and for controlling the rapidly increasing world population were enormous. And indeed, in the more than twenty years that it has been widely used, the Pill has made a major impact on individual lives, families, nations, and the world. "Modern woman is at last free, as a man is free," Clare Boothe Luce declared, "to dispose of her own body, to earn her living, to pursue the improvement of her mind, to try a successful career." Over fifty million women now use the Pill worldwide.

The public euphoria associated with the easy availability of this simple and effective birth-control device and the high hopes demographers, medical scientists, and political leaders attached to its use in controlling population growth would last for nearly a decade. But it was eventually diminished by disturbing reports of fatal vascular side-effects.

The first such report appeared in the *Lancet* in 1961. It came from a general practitioner in Suffolk, England, who was treating a woman patient with an oral contraceptive for endometriosis, a disease of the womb. The patient developed a blood clot (thromboembolism) in her lung. Following this report, many others appeared—so many, in fact, that by 1963 it was abundantly clear that careful epidemiologic studies were needed to determine whether these blood clots were merely coincidental, or whether they were a direct effect of the Pill.

Soon thereafter, research on the relationship between blood clots and the Pill was expanded to other areas, such as heart attacks, strokes, and high blood pressure, and the interplay between the Pill and other risk factors, such as age, smoking, and diabetes.

The only tool available for providing relatively quick answers is the *case-control study*. With this technique, women with blood clots who had used the Pill were compared to a control group of others of the same age, socioeconomic background, and general state of health who had never used it. By the late 1960s a number of case-control studies had produced strong evidence that women using the Pill had a greater risk of developing blood clots than nonusers. However, different studies came up with widely different risk data for a first episode of blood clot, ranging from three to eleven times greater in Pill users than in nonusers.

The public was understandably confused by this range of figures. A threefold greater risk is one thing, an elevenfold greater risk quite another. Yet the risk of blood clots in Pill nonusers was so low that some scientists argued that a threefold or even an elevenfold increase in risk was just not significant. Others didn't agree. Meanwhile, the FDA, most leading physicians, and Planned Parenthood continued to hold that the Pill was safe.

Some observers who believe that oral contraceptives are not safe now feel that the Pill never underwent fair scrutiny. A lot was riding on it—the "freeing" of women, for one thing; population control for another. (Margaret Sanger, a champion of the diaphragm, once wrote that the future of civilization depended upon "a simple, cheap, safe contraceptive to be used in poverty-stricken slums and jungles, and among the most ignorant people.") "No wonder," writes Barbara Seaman, cofounder of the National Women's Health Network, that the Pill "long enjoyed diplomatic immunity from criticism."

The most one could deduce from these case-control studies was that while Pill users faced an increased risk of having blood clots, the magnitude of this risk varied. But neither medical scientists nor the general public liked dealing with a variety of possibilities on such a critical health issue, and so a number of *cohort studies*—in which a group of people are monitored for an extended period for the development of a symptom, a disease, or a complication—were begun in 1968 to help pinpoint the risk level. Two of these studies were started in Great Britain and one in the United States. A total of eighty thousand women, half of whom took the Pill, participated.

While the results of the studies were being awaited, the press belatedly began covering the growing controversy over the safety of oral contraceptives. In January 1970, Senator Gaylord Nelson held widely publicized hearings on this subject. Witnesses suggested that the Pill had not been adequately tested prior to its approval, and that Pill users had not been adequately informed of the risks of its use. The FDA and the medical establishment continued to call oral contraceptives safe, but in the following months an estimated 19 percent of all Pill users quit, at least temporarily. Some Pill-boosters claimed that one hundred thousand unwanted pregnancies resulted, and called the products of these births "Nelson Babies."

By 1974 the three cohort studies began yielding a wealth of information about the risks associated with the Pill. Pill use had peaked in the United States, with 25 percent of women aged fourteen to forty-four years taking some form of oral contraception. (By 1982 fewer than 15 percent—ten million American women—were on the Pill.)

The cohort studies, like the case-control efforts, did not produce identical results, but the results were not significantly different in a statistical sense. Taken together, they showed that the risk of blood clots and other cardiovascular complications of Pill use (strokes and heart attacks) was significant but lower than that found in case-control studies. The Population Information Program of the Johns Hopkins University thinks this is due to three reasons: (1) more observations have led to more precise data; (2) women with a history of circulatory problems or with characteristics identified as risk factors in previous studies have stopped using the Pill; and (3) lower-dose pills, associated with fewer side-effects, are now more commonly used.

In the cohort studies done in Great Britain, it was found that 19 of 10,000 women per year who used the Pill developed blood clots in superficial leg veins. Among nonusers of the Pill, 8 women per year out of 10,000 also developed clots. The ratio of 19 per 10,000 to 8 per 10,000 is slightly greater than 2, and this is known as the "relative" risk. It indicated that Pill users were slightly more than twice as likely to develop blood clots as nonusers.

But people also wanted to know the "attributable" risk—the additional risk of using the Pill over and above that of not using it. This is essentially the difference in the incidence of ill-effects between users and nonusers. In this case, the attributable risk— the difference between 19 and 8—was 11, a significant "extra" 11 cases of blood clots per 10,000 users per year, of great importance to medical scientists.

Additionally, the risk of blood clots forming in leg veins, which increases after surgery, was greatly increased for Pill users. Sixty-one of 10,000 Pill users developed postoperative blood clots within a year after having surgical operations, while only 30 of 10,000 nonusers did, but since 1968, the British cohort studies have found only five fatalities associated with these clots.

Recent research has thrown much light on how the Pill causes blood clots. Oral contraceptives contain two basic chemical ingredients—the hormones estrogen and progestin. The estrogen component (of which there are actually many different kinds) has been found to increase the clotting ability of the blood, while both components seem to slow the rate of blood flow through the veins and cause changes in the linings of the arteries and veins that favor blood-clot formation. Studies also reveal a direct relationship between the estrogen content of a contraceptive pill and the risk of blood clots among its users: pills containing lower concentrations of estrogen cause only one-third to one-half as many blood clots as those containing higher concentrations. This finding has now led to the development of low-dose oral contraceptives that retain enough potency to be effective, but which cause fewer side-effects, such as clots. But nearly half of the Pill users in the United States continue to take pills with high levels of progestin and about one in eight are still using pills containing high doses of estrogen.

Recent research by Dr. Bruce Stadel, however, has shown that there is also much variation in the levels of estrogen in the blood of women taking the Pill. This may be due to differences in gastrointestinal absorption of the Pill's contents, the distribution of these contents in the body, and their rate of elimination. It suggests, says Stadel, that pills containing what are considered to be "safe" estrogen levels may be safe for some women but not for others: two women taking the same pill might end up with different estrogen concentrations in their blood and thus with different risks of developing blood clots.

Another factor now thought to play a role in blood-clot formation among Pill users is their blood type. Women with blood types A, B, or AB normally have a twofold greater risk of developing deep leg-vein blood clots than women with type O blood.

As the evidence concerning estrogen levels and blood type shows, a broad series of circumstances can affect the degree of risk for the user of oral contraceptives.

A number of case-control and cohort studies also focused on other risks associated with use of the Pill, such as heart attacks and strokes. Either blood clots or fat deposits (atherosclerosis) can block the arteries that feed the heart, and when the blood supply to a portion of the heart muscle is greatly reduced or stopped, a heart attack (myocardial infarction) results. Several studies have made it clear that the risk of heart attacks among Pill users aged twenty-five to forty-nine years is about four times greater than among nonusers. (This increased risk seems to persist for ten years after going off the Pill.) A number of British and American studies have also shown that the Pill multiplies the naturally increased risk of myocardial infarction that accompanies aging in the general population.

Besides this, a number of studies have now shown that strokes are concentrated in older Pill users who smoke and who have high blood pressure. But these studies have come up with different magnitudes of risk for women taking the Pill who have one or any combination of these characteristics. Some of these studies have shown that the risk of hemorrhagic stroke (caused by a ruptured blood vessel) lingers even after Pill use stops; other studies do not show this. One study published in the Lancet in 1978 reported that the risk of hemorrhagic stroke may increase with the duration of Pill use.

Today, approximately 450 women die each year in the United States because of the effect of the Pill on their hearts and blood vessels. Studies conducted by the Royal College of General Practitioners in Britain and by the Centers for Disease Control in the United States have shown that half of these deaths could be avoided if women who use the Pill did not smoke. It is also clear that women over the age of forty-five have the greatest risks from using the Pill. Concerning these women, the Royal College has said, "We believe that use of oral contraceptives can be justified only in exceptional circumstances." Pill use appears safest in women under age thirty-five who do not smoke and who do not have other risk factors such as high blood pressure.

On October 22, 1983, two new research reports appeared in the *Lancet* linking the Pill to cancer causation. A case-control study of 314 women with breast cancer and 314 controls in Los Angeles detected a significant risk of breast cancer in women who used pills with high progestin content for long periods of time beginning before they were twenty-five years of age. A study in Great Britain, published in the same issue of the *Lancet*, reported a strong link between the use of the Pill and risk of cervical cancer. Four days after the study results were published, the British government's Committee on the Safety of Medicines recommended that women be prescribed a pill with the lowest suitable content of estrogen and progestin. Some consumers in the United States urged the FDA to take a similar step, or even consider banning high-dose pills.

However, in May 1984 it was announced that the national Centers for Disease Control and Boston University's Drug Epidemiology Study were unable to substantiate the Los Angeles study. The FDA's advisory committee on oral contraceptives voted unanimously to reject the findings of the Los Angeles study. Dr. William Filler, Jr., a member of the FDA's committee, said, "The tragedy of this study is that it was so widely publicized before the scientific community had an opportunity to dissect it thoroughly and reveal its flaws." Dr. Solomon Sobel, director of the FDA's division of metabolism and endocrine drugs, said, "We don't feel the link between early usage of oral contraceptives and the development of breast cancer has been established."

The FDA committee that rejected the Los Angeles study did, however, recommend that Pill prescription labels advise women using the Pill to get annual breast examinations along with Pap smears to check for cancer of the cervix.

The five researchers from the University of Southern California stood by their findings. Dr. Malcolm Pike, chief author of the study and now director of the Imperial Cancer Research Fund's epidemiology unit in Oxford, England, told *The New York Times*, "I haven't changed my opinion" about the link between the Pill and breast cancer.

From all that we have seen, it is clear that defining the acceptability of risks associated with the Pill is not an easy matter. Although some rough predictive value may be assigned to certain risks based on extensive studies of large groups of people, there is no sure way for any individual user of oral contraceptives to know whether or not she will ever become a "risk victim." Nor is there any predictive certainty that someone who takes the risks will derive all or any of the benefits. On the other hand, various risk factors—from blood type to age—can now be considered in determining whether to take or shun the Pill as a birth-control device.

In the final analysis, women considering use of the Pill must weigh its known risks against its known benefits, quantifying each as much as possible. The greater the benefits and the greater their certainty, the more acceptable the risk becomes. For one thing, the risks and benefits of Pill use must be compared to those associated with other forms of contraception. Diaphragms and condoms, for example, may be intrusive, and not entirely reliable, while intrauterine devices (IUDs) and the new "sponge" contraceptives pose hazards of their own.

The benefits ranked on the side of the Pill include the almost certain avoidance of pregnancy, which carries risks, and avoidance of abortions and the enormous social, economic, and emotional problems associated with unwanted children. To these have now been added a number of both short- and long-term health benefits, such as the Pill's favorable impact in the areas of rheumatoid arthritis, cancer of the ovaries, and pelvic inflammation. For when all is said and done, defining the risks, not the benefits, is the critical element in this equation, for the risks in the case of the Pill are indeed serious. Despite the application of modern medical research tools and sophisticated mathematical models, it is still a lot easier to describe these risks than to define them fully.

Close-Up (2)
TOXIC SHOCK SYNDROME

"I went dancing the night before in a black velvet Paris gown, on one of those evenings that was the glamour of New York epitomized. I was blissfully asleep by 3:00 a.m. Twenty-four hours later, I lay dying, my fingers and legs darkening with gangrene. . . ."

—Nan Robertson, in *The New York Times Magazine*

Nan Robertson, a reporter for *The New York Times*, spent ten and a half weeks in the hospital and underwent several operations that left her with eight partially amputated fingers, a devastating disability for someone in her profession. When she left the hospital she couldn't turn a door knob, wash or dress herself, button a blouse, or tie her shoelaces. With tenacious courage she is now able to do most of these things, and type as well.

Others, however, haven't been so lucky. They have died within twenty-four hours of developing the symptoms that accompany serious cases of toxic shock syndrome (TSS). One of the reasons Nan Robertson survived was that her doctor immediately recognized her affliction. She had four of the five early, acute symptoms of TSS—diarrhea, vomiting, a rash, and falling blood pressure. Eventually she developed the fifth, a fever of over 102 degrees Fahrenheit. We now know that all of these symptoms are caused by the toxins of a bacterium known as *Staphylococcus aureus*.

Prior to June 1980, however, not many physicians had ever heard of toxic shock syndrome. Faced with patients displaying these symptoms, they were baffled. Finding out what was wrong took time—precious hours these patients couldn't afford. The fatality rate for TSS is about 8.4 percent of all reported cases, and treatment was often a matter of guesswork or based on incomplete information from laboratory tests. Not surprisingly, several dozen of these patients died, often without their physicians knowing exactly why. By the end of 1983 over one hundred TSS victims had died.

TSS was first described in 1978 by Dr. James Todd, director of Infectious Diseases at the Denver Children's Hospital, and several colleagues in an article published in the *Lancet*. Although the journal is widely read by physicians in the United States, Todd's article did not attract much attention. In it, he described seven cases, six among girls and one in a boy, aged eight through seventeen years. Three of the girls had been menstruating and using tampons when they became ill, but Todd and his colleagues didn't draw an association between the disease and menstruation. They did demonstrate, however, that the disease was caused by a toxin produced by the *Staphylococcus aureus* bacterium. (Although all of Todd's patients had severe forms of the disease, it is now known that some cases may be relatively mild and go undetected.)

Nothing more happened until a year later, when a perceptive epidemiologist, Dr. Jeffrey P. Davis of the Wisconsin Division of Health, noticed that three cases of TSS had been reported to the state health department in the month of November 1979. He decided to investigate these cases further, in the hope of finding out how and why people became infected, and where in the body the infecting bacteria grew.

Because all three of the Wisconsin cases had occurred in women, Davis began looking at characteristics unique to these three women. He found that the three women were menstruating at the time they became ill, but he did not know what relationship—if any—this had to TSS.

In order to get more information, Davis launched surveillance programs in Wisconsin, and another was begun in Minnesota. Every doctor in both states was sent a letter telling them about the disease, and indicating that it might in some way be related to menstruation. The doctors were asked to immediately report any case of TSS to public health officials.

By late January 1980 Davis had received twelve case reports, all involving women; eleven had been menstruating when they became ill. Davis had superb investigative instincts. Because he knew that menstruation, a normal bodily function in women, couldn't cause TSS, he interviewed the seven Wisconsin victims among the twelve reported cases, hoping to uncover something the case investigations had missed. In so doing he was exploring unmapped terrain.

His detective instincts paid off. He found out that most of the women had used tampons while menstruating. Although he couldn't yet prove it, Davis suspected that tampon use was in some way related to TSS.

Davis and his colleagues in Minnesota reported their findings to the Centers for Disease Control in Atlanta, Georgia. In February 1980 the center assigned Dr. Kathryn Shands, an Epidemic Intelligence Service officer, to investigate the problem. Meanwhile, Davis initiated a case-control study, a good epidemiological tool for quickly defining risks. While Davis was conducting this study, Shands contacted Dr. James Todd in Denver, who by now had collected a total of thirty-five case reports of TSS, ten of them in males. She attempted to pinpoint common characteristics in these cases, but the ten male cases of the disease made her wonder about Davis' hunch that tampon use was in some way to blame.

Several medical investigators now began to advance on TSS on a number of different fronts. In March 1980 Dr. Christian Schrock of the Minnesota Department of Health described three cases of the disease in a letter to the editor of the *Journal of the American Medical Association*. He concluded that the disease seemed to occur only in women, at an age of active menstruation and at or near the time of menstruation.

But while these conclusions seemed true on the basis of Schrock's own three cases, there remained the matter of James Todd's ten male cases in Denver. Obviously menstruation and tampon use were not the only factors, and as everyone involved in the investigations knew, they might be nothing more than coincidental factors.

In late May 1980 Davis completed his case-control study and reported the results to the Centers for Disease Control. His data suggested an association between tampon use and TSS, as did the results of a case-control study done in Utah. Still, these results weren't convincing beyond the shadow of a doubt. The haunting question about tampons and TSS—whether they had a cause-and-effect relationship or were just coincidental—remained unanswered.

By this time, the Centers for Disease Control had received fifty-five more case reports of TSS from around the country. These were announced in the May 23, 1980, edition of the Centers' *Morbidity and Mortality Weekly Report*. The Centers avoided describing diagnostic criteria for the disease, but did stress that it often occurred in menstruating women.

But at this point, there still remained many things that were unknown about toxic shock, especially any firm evidence about how people contracted it. However, the Centers' announcement did have several positive results; for one thing, it led to extensive coverage in the media, which made both physicians and laymen acutely aware of the disease. This, in turn, led to the more rapid and accurate diagnosis of cases of TSS, which in turn gave the Centers increasing data on the disease. And the more epidemiologists learned about it from the ever-enlarging number of cases, the more they felt they would decipher how it developed.

The national publicity surrounding the Centers' May 1980 announcement resulted in a Congressional hearing in which Dr. Shands and Dr. William Foege, the director of the Centers for Disease Control, outlined the case-control study they would conduct from June 13 through June 19, 1980. Just as that study got under way, however, Dr. John Bennett, the head of the Centers' Bacterial Disease Division, was reaching the same conclusion that Dr. Jeffrey Davis had reached in Wisconsin several months earlier: The case reports coming into the Centers from around the country suggested a relationship between tampon use and TSS.

The Centers conducted a seven-day case-control study as planned; this consisted of questioning, by telephone, fifty-two women who had had TSS (the women included in the study were required to have had diarrhea, vomiting, low blood pressure, a fever over 102 degrees Fahrenheit, and a skin rash). Fifty-two other women who had not had TSS served as the controls. Both groups of women were asked detailed questions about their contraceptive use, marital status, frequency of sexual intercourse, frequency of sexual intercourse during menstruation, and brand of tampon or sanitary napkin used.

The published results caused a national sensation. On June 27, 1980, the *Morbidity and Mortality Weekly Report* announced that the association between toxic shock syndrome and tampon use was statistically significant. The Centers for Disease Control also announced that the results of Dr. Jeffrey Davis' Wisconsin study showed a statistically significant association between tampon use and the disease, although the results of the Utah study were not statistically significant. None of the three studies had determined the degree of risk for getting TSS from various tampon brands or from highly absorbent tampons, and none had answered the question of how and why males got the disease. But the Centers for Disease Control recommended that women who had had the disease not use tampons for at least several menstrual cycles after their recovery.

Although it was devastating for those who got it, the studies showed that the risk of getting TSS was very low. Thus the Centers did not suggest that all women stop using tampons. However, the Centers' medical scientists did suggest using tampons only intermittently during menstrual cycles in order to minimize the risk of TSS. All of this hedging advice simply reflected the fact that there was still much that remained unknown about the disease and how to prevent it.

During the summer of 1980, attention was riveted on determining the degree of risk for TSS among tampon users. The almost daily publicity produced an enormous increase in the number of cases being reported directly to the Centers for Disease Control by physicians, patients, and the relatives and friends of patients. By early September 1980 the Centers had received reports of 272 cases, which amounted to an epidemic of TSS. All of these case reports had to be investigated, however, since not all were truly cases of TSS.

From September 5 through September 8, 1980, the Centers conducted a case-control study aimed at defining the risk of TSS with various brands of tampons. This study, whose results were published in the *Morbidity and Mortality Weekly Report* of September 19, 1980, showed that tampon users had a tenfold greater chance of getting the disease than did nonusers. The highest level of risk was associated with superabsorbent Rely brand tampons, which—although they claimed only 20 percent of the market—were associated with 70 percent of TSS cases.

The Centers' study also threw more light on the role of *Staphylococcus aureus* in TSS. The organism was found in the vaginas of 98 percent of the women who had the disease. In cases involving males and in other cases not associated with menstruation, the organism was found in the skin, bones, and lungs. All of the *Staphylococcus aureus* bacteria found in toxic shock patients were resistant to penicillin.

On September 22, 1980, the Procter and Gamble Company, while disputing that there was any link between its Rely tampons and TSS, voluntarily withdrew them from the market. Shortly thereafter, the company found itself the object of several hundred lawsuits, many of which are still in process.

At about the same time, however, a number of medical experts began combing back over all of the TSS studies to check their statistical accuracy, and some of these researchers claimed they found biases and defects that cast doubts on the studies' results.

Among those who questioned the validity of the Centers for Disease Control studies was Dr. James Todd, who had first described TSS in 1978. Testifying for Procter and Gamble in a civil suit in Iowa, Todd said that doctors' biases about toxic shock and tampon use affected the results of the Centers' study linking Rely tampons to the disease. Todd cogently observed that the publicity caused by that study had made doctors who were treating menstruating women complaining of nausea and fever—two symptoms of toxic shock—falsely assume that these women had a minor form of the disease. Thus a number of supposed cases of TSS among Rely users may not truly have been TSS. And by inaccurately counting such cases as toxic shock, Todd charged, doctors participating in the studies may have given the Centers' researchers an exaggerated notion of the risks posed by tampon use. (An Iowa jury didn't buy this idea, awarding $300,000 to the family of a woman who died from TSS.)

Todd was not the only one to express doubt about the Centers' studies. In the August 1982 *Journal of the American Medical Association* three Yale University researchers critically analyzed all of the studies that had linked TSS and tampon use.

Coming from respected scholars not in the employ of industry, their negative criticism, backed by factual and statistical data, could not be dismissed.

More recent research, however, has supported evidence linking tampons to TSS. In early 1982 two microbiologists working independently in New York City produced some convincing evidence that Rely tampons were associated with a large proportion of toxic shock cases. Dr. Philip Tierno, a microbiologist at New York University Medical Center in Manhattan, took the superabsorbent material (known scientifically as carboxymethylcellulose) from these tampons and placed it in test tubes. When he added bacteria normally found in the vaginas of menstruating women, the bacteria broke down the carboxymethylcellulose into a glucose gel, an important food source for *Staphylococcus aureus*.

At Manhattan's nearby Veterans Administration Hospital, Dr. Bruce Hanna, a microbiologist, essentially confirmed Dr. Tierno's work and also isolated the bacterial enzyme that breaks down the carboxymethylcellulose. When the broken-down material was inoculated with *Staphylococcus aureus* the organisms grew at a very rapid rate. Dr. Hanna also found that although other tampons contained carboxymethylcellulose, the risk of Rely tampons causing TSS may have risen because when moistened these particular tampons swell, with the superabsorbent material turning into a gel that traps *Staphylococcus aureus*, which then produces its toxins; the gel also serves as a protective barrier from human immune mechanisms and bacteria-destroying white blood cells. If Dr. Hanna is right, then Rely tampons both provided a food source for *Staphylococcus aureus* and shielded them against destruction by the human body.

Although they have not yet been duplicated, these research results, and others, seem to indicate that the *Staphylococcus* bacterium grows better, produces its toxin more efficiently, or both, in the presence of menstrual fluid and tampon material. As a result, tampon manufacturers have largely stopped using carboxymethylcellulose and have substituted cotton and rayon instead.

Some researchers have challenged the purported mechanism for the development of TSS. Dr. Patrick Schlievert of the University of Minnesota and others, while agreeing that the toxic shock bacteria grow well on degraded Rely tampons, point out that the glucose generated by the tampon's breakdown *prevents* the bacteria from producing their toxin. They think that the *Staphylococcus* bacteria instead multiply to dangerously high levels in the vagina and produce the toxin in an environment made favorable by oxygen that is introduced with the tampon.

Schlievert has also proposed a mechanism for how TSS develops. He suggests that the toxin damages both the human immune system, crippling its fight against infection, and the liver, preventing it from breaking down the toxin as the blood carries it through that organ.

Extensive research has also confirmed what medical scientists have known for a long time—that toxin-producing *Staphylococcus aureus* can be found at any given time in the nasal and throat passages, on the skin, and in the vagina in millions of women. But the research has come up with a new finding—5 percent of menstruating women carry the organism in the vagina. Since most persons who carry the bacteria are immune to the effects of the toxin, however, only a small percentage come down with TSS. Male carriers who are not immune can come down with TSS from bacteria in their nasal passages, as can women in whom the bacteria produce their toxin in the vagina (and most reported cases to date have been observed among women). Persons of either sex who are incompletely immune tend to have mild cases of TSS. The risk is obviously greatest for menstruating women using superabsorbent tampons, who account for 85 percent of all cases of TSS among women.

Because there is strong epidemiologic evidence for an association between tampon use and TSS, a special panel convened by the Institute of Medicine of the National Academy of Sciences recommended that women minimize their use of high-absorbency tampons. And, said the committee chairman, Dr. Sheldon M. Wolff of the Tufts University School of Medicine, "minimize" meant that most women shouldn't use them.

In June 1982 the Food and Drug Administration announced that by December of that year a brief warning would have to appear on tampon boxes, with a longer explanation on leaflets inside. The warning read, in part: "Attention—Tampons are associated with toxic shock syndrome." Most manufacturers of tampons had already begun printing warnings on their boxes long before this, but there was no consistency to this practice. And two years after the intense publicity about tampons and toxic shock had died down, manufacturers—including the makers of superabsorbent tampons—were advertising their products as vigorously as ever.

The number of toxic shock cases reported to the Centers for Disease Control has dropped off drastically—from 859 in 1980 and 552 in 1981 to about 35 cases per month today. This has encouraged women, perhaps overeagerly, to return to the carefree use of tampons. In *The New York Times* health columnist Jane Brody reported: "Public concern about toxic shock syndrome has waned and tampon use has returned to former levels."

This confidence, however, may be unwarranted. The most intensive search for cases of TSS, which took place in Minnesota in 1980 and 1981, turned up 9 cases per 100,000 menstruating women, which meant that there should be at least 4,500 cases of TSS occurring nationwide each year—which, with today's 35 or so cases per month, means that there are more than 4,000 unreported cases annually. The Minnesota study also found no slackening in the number of toxic shock cases *after* Rely-brand tampons were taken off the market. When that happened, reported Minnesota state epidemiologist Dr. Michael T. Osterholm, women switched to other superabsorbent brands and the disease rate remained the same.

Even the results of the first wave of lawsuits brought against tampon manufacturers by alleged victims of TSS did not stem the industry's, or the public's, enthusiasm for tampons. In the most dramatic of these lawsuits, Lynette West, a twenty-two-year-old Los Gatos, California, woman who had been using Johnson and Johnson's O.B. tampons when she became ill, was awarded $10.5 million by a superior court jury in Santa Clara County. The size of the award, by far the largest any TSS plaintiff had won, was considered surprising because Ms. West had not died from TSS, and in fact seemed to have completely recovered.

When Procter and Gamble voluntarily withdrew Rely tampons from the market in September 1980 (the decision to jettison Rely wasn't difficult, since it accounted for only 1 percent of the company's revenues), it set up a toxic shock scientific task force of its own. Through it, Procter and Gamble spent $3 million on toxic shock research.

Among those funded by Procter and Gamble was Merlin Bergdoll, Ph.D., a professor at the Food Research Institute of the University of Wisconsin. Bergdoll became well known for his studies of TSS. In the May 9, 1981, issue of the *Lancet*, he and his associates reported the isolation of a protein toxin from *Staphylococcus aureus*. It was found in 94 percent of cultures from TSS patients, but only 12 percent of cultures from women without TSS showed it. (A similar toxin had been isolated at about the same time by Dr. Schlievert. Most experts now agree that these two toxins are identical and refer to it as toxic shock toxin or TST. Other toxins have also been isolated in victims of TSS and some experts think that a combination of them and TST cause the disease.)

Bergdoll and his group also uncovered other new information crucial to an understanding of toxic shock. In testing women for antibodies (protective substances in the blood) against TST they found that 95 percent of women over thirty years of age were protected. This explained why the greatest proportion of cases of toxic shock are seen in very young women. The age of the average case is 18.8 years. What all of this means is that as women get older they become immune to the toxic shock toxin.

Although Bergdoll shared many of his research findings with the medical community, there were others he did not. And Procter and Gamble, the financial sponsor for much of his research, went to great lengths to hold back publication of some of Bergdoll's research results. This effort ended during a trial in Fort Worth, Texas, in which Procter and Gamble was being sued by a TSS victim who had used Rely tampons.

Tammy Lynn Wallace, a twenty-six-year-old Texan who was hospitalized in 1979 for twenty-seven days for TSS, alleged that she contracted the disease from using Rely tampons. Her lawsuit against Procter and Gamble came to trial in November 1983, during which her attorney was successful in forcing Procter and Gamble to hand over research findings that Bergdoll had produced back in 1981. Those findings (still characterized by Procter and Gamble as unconvincing in establishing a cause and effect relationship between Rely tampons and TSS) struck many as a crucial piece in the TSS puzzle.

Through legal maneuvering Procter and Gamble was able to hold up publication of this data. Under the doctrine of "attorney's work product," a lawyer cannot obtain the opposing side's work and use it. If the lawyer wants this evidence, he/she has to create it independently. Therefore Procter and Gamble was able to hide research findings that many now recognize as damaging to its defense. This is no small matter because as of January 1984, two hundred lawsuits had been pressed against the company, four of which had gone to trial. (The vast majority were being settled out of court.)

Bergdoll's data finally came out in the Fort Worth courtroom. It essentially showed that toxin-producing strains of *Staphylococcus aureus* produce much more enterotoxin F (SEF), now known as TST, when grown on Rely super tampons than on most other tampons. But that was not all. Michael Liles, Wallace's attorney, also told the court that the research had shown that the polyester foam cubes (a unique feature of Rely) are a major source of toxin production.

These revelations were not the only ones to emerge at the trial. More startling still was the admission by a Procter and Gamble scientist that the company had independently reproduced and confirmed Bergdoll's results in mid-1982.

This information impressed the medical world because it provided laboratory confirmation of what epidemiologists had already suspected—that TSS was more related to the absorbency of tampons than to their frequency of use. This critical clue had surfaced in 1981 in the Tri-State Toxic Shock Syndrome Study (Minnesota, Wisconsin, and Iowa). But laboratory confirmation was lacking. In the view of some, Bergdoll and Procter and Gamble have now provided confirmation.

Prior to the Fort Worth court trial, Bergdoll said that he was not going to publish the results of his research because no hard conclusions could be drawn from it. As late as March 1984 he still maintained that his study findings represented preliminary raw data and that they were a very small part of his work with TSS.

The Fort Worth case ended in a mistrial when the plaintiff's attorney became ill. A spokesperson for Procter and Gamble said, "There's no scientific proof that tampons are linked to TSS."

Attempts to define the risks of TSS with tampon use have spanned several years. During this time scientists have had to contend with less than perfect epidemiologic tools for measuring risk, the doubts expressed by some respected scientists, the purposeful suppression of crucial laboratory evidence by a company in jeopardy, and the long time periods required for producing and confirming laboratory and clinical investigations.

All of this illustrates that defining risks is no easy matter. If called upon to comment on the risks of TSS with tampon use most experts would say: We're sure, or almost sure, but perhaps not quite sure.

Close-Up (3)
X RAYS

"It might be better to take everybody's blood pressure than [to do a] chest X ray."

—Dr. Reginald F. Brown,
American College of Radiology

For years people had routine chest X rays without giving them a second thought. Today some people are refusing to have them at all, while others are keeping their own personal records of radiation exposure, demanding lead shields when they are X-rayed, and asking about the dosage they'll receive before they even make an appointment for an X ray.

Why all this concern about low-dose radiation? After all, almost 50 percent of the radiation to which we are exposed comes from natural sources—cosmic rays and minerals in the ground. Radiation turns up in food, water, and air. There is certainly nothing we can do to protect ourselves from these radiation sources. And obviously our ancestors lived with this type of radiation without observing any significant ill-effects.

However, this does not mean that the radiation we are exposed to simply by living on earth is necessarily benign. Some scientists now estimate that in the United States natural background radiation may cause as many as forty-five thousand cancer deaths each year; others believe that this figure is much too high.

The average person receives about 100 millirems of "natural" radiation a year. (A millirem is one-thousandth of a rem, which stands for *roentgen equivalent man* and refers to the amount of radiation needed to produce a particular amount of damage in living tissue.) According to standards set by the United States government, a person who works with ionizing radiation, such as a nuclear power-plant employee or X-ray technician, should get no more than 5 rems of radiation a year. The suggested standard for everyone else is lower—.170 rems, or 170 millirems.

Not everyone, however, receives the same low yearly dose of radiation that most people receive. Smokers take polonium-210, from tobacco plants, directly into their lungs. People who live in brick houses receive more gamma rays than those who live in wooden houses (radiation is given off by bricks, cinderblock, concrete, and other building materials); there is growing concern about radon gas trapped inside such homes that have been made more airtight to conserve energy. As many as 10,000 lung cancer deaths in this country may be caused by radon gas, declared a recent report in *The New England Journal of Medicine*. (According to some estimates the average American stands a one-in-two-hundred chance of dying due to exposure to radon gas.) Even such everyday consumer products as color TV sets, some smoke detectors, the luminescent dials of clocks and watches (which contain radium), and the glass used in eyeglasses produce radiation.

The mining and processing of certain materials, such as uranium oxide, causes about 2.5 percent of all human exposure to radiation (the proportion is, naturally, higher for miners and those who live near uranium mines and processing plants). Although underground weapons tests in the United States and the Soviet Union produce little atmospheric radiation today, radioactive fallout from nuclear weapons tests still constitutes about 3 percent of the background exposure to radiation, since long-lived strontium-90 and plutonium isotopes from earlier tests are still with us, and France and China continue to conduct tests in the atmosphere.

Additionally, everyone on earth receives at least 30 millirems of radiation annually from cosmic rays, although people who live at high altitudes, where the protection provided by the atmosphere is reduced, receive more. As a result, people who live in the Rocky Mountain region may get more than 100 millirems of cosmic radiation each year (the equivalent of one chest X ray), the crews of jet airliners receive as much radiation a year as many workers in nuclear power plants, and a passenger who takes a five-hour transcontinental flight gets a dose of 3 to 4 millirems. (This comprises a cancer risk, according to one estimate, of five in ten million per flight.)

Radiation damages the cells of the body in two principal ways. At extremely high doses, it kills the body's cells. At lower but still significant doses, it causes cells to become cancerous. Paradoxically, the cell-killing ability of radiation is sometimes used for treating cancers that have been caused by radiation in the first place. Radiation can also damage the chromosomes of a cell and the genes these chromosomes contain, resulting in mutations that are seen in subsequent generations of the cell or the larger organism to which it belongs.

Because radiation is both good and bad, depending upon a number of variables, it was obvious from the time X rays were first widely used, in the late nineteenth century, that exposure to it had to be measured and monitored in some way. For a number of years scientists have tried to define safe levels of radiation with little success. The question always comes down to: How little is too much? Many believe that there is no "threshold" dose of radiation below which the risk of cancer is entirely absent; others think that reasonable estimates can be made. A new body of thought, called hormesis, even holds that a little bit of radiation may actually be beneficial. Controversial and widely divergent opinions about the relationship between radiation dose and cancer, the long period many cancers take to develop, and the confounding influence of other cancer-causing factors in the environment make it extremely difficult to define this risk.

This problem is most pressing in the area of medical and dental X rays, which produce 40 percent of our total radiation exposure (and 90 percent of the manmade variety). During 1980, for example, about 86 million people in the United States had dental X rays and 161 million received medical X rays. Opinions vary widely on the dangers posed by these procedures. One leading expert has charged that in the next thirty years medicine will in effect sign about 1,400,000 death warrants because of unnecessary radiation exposure. But another expert, Dr. Edward W. Webster, chief of radiological sciences at Massachusetts General Hospital, objects to this. "To say that we're signing death warrants in hospitals," he comments, "is an abuse of what we precisely know about low-level radiation."

What we precisely know is, unfortunately, quite little.

X rays were first discovered in 1895 by Wilhelm Roentgen, a German physicist. His discovery would not have gone very far were it not for E. H. Grubbe, an established Chicago manufacturer of high-voltage tubes who promoted the therapeutic uses of radiation and made a handsome profit doing so.

In the following year, 1896, Thomas A. Edison invented the fluoroscope, which is still used today to visualize the body's anatomy in motion. Because modern medicine had precious little to offer people, it is no surprise that X rays were used for treating everything from acne to fibroid tumors of the uterus. Patients were blasted from head to toe with the rays, as were most of the physicians who used them and who had little or no training in their use and almost no knowledge of their dangers.

The immediate effects of excessive radiation—burns and skin ulcers—became known almost at once. So did the relationship between radiation and cancer. By the turn of the century the medical literature carried numerous case reports of radiation-induced cancers. Most of these occurred in the operators of X-ray machines and in X-ray researchers. Among the latter was

Clarence Dally, the first American X-ray experimenter to die from the effects of radiation. Dally, who died in 1904 of mediastinal cancer after having both arms amputated, had been an assistant to Edison and had helped develop the fluoroscope. Edison regularly fluoroscoped Dally during experiments and Dally received additional radiation exposure during experiments he himself conducted.

The concern for radiation damage in these early years focused mainly on X-ray operators, who were perceived as being at the greatest risk. Finally, in 1916, the British Roentgen Society published standards for the use of X rays, but these were of an advisory nature, and not binding on anyone. By then X rays were being widely used by physicians, many of whom still had little or no training in their use, which is to some extent still true today.

Insurance companies, quick to recognize risky occupations, began classifying physicians who worked with X rays as a high-risk group, and charged them higher premiums. This, however, had little effect on X-ray practices. By 1927, a survey of physicians using the rays revealed that 36 percent of their marriages were childless, and that the rate of birth defects in their children increased from 2.6 percent before they personally used X rays to 4 percent afterward. Subsequent studies showed that these physicians also had significantly higher than normal death rates from leukemia and cancer. But even these dramatic findings did not produce a diminution in X-ray use. Physicians using the rays, like the radiologists of today, defended the benefits of their use and pointed out that improved techniques requiring lower X-ray doses made risk assessments irrelevant.

At first glance this might seem self-destructive, but it must be remembered that until the 1950s, physicians continued to use X rays widely, for treating everything from ringworm to adenoids. The financial profits from this practice were enormous, while the terrible consequences for some doctors and some of their patients lay far off in the future—sometimes as far as twenty years—and were often viewed as purely speculative.

One example of this abuse of X rays was the consistent use of fluoroscopy on tuberculosis patients in the 1930s and 1940s.

During this era, before effective drugs for the disease became available, lungs affected with tuberculosis were often collapsed to prevent spread of the infection to other parts of the patient's body. These collapsed lungs were regularly monitored by fluoroscopy. The long-term result was that a significant proportion of women subjected to these X rays later developed breast cancer.

By the 1950s a number of studies had clearly shown that high-dose radiation caused everything from cancer to congenital defects in unborn babies. Improved technology, radiologists' and physicians' adherence to new guidelines, and the fear of devastating malpractice suits curtailed much of the unnecessary exposure to high-dose radiation. Almost predictably, the focus of concern then shifted to low-dose radiation.

In 1972 a National Academy of Sciences/National Research Council Advisory Committee on the Biological Effects of Ionizing Radiation (BEIR) published a report that changed attitudes toward the risks of low-dose radiation. The report said that on the basis of then-current knowledge, there appeared to be no threshold above which cancer was induced and below which it wasn't. The committee, while admitting that its information was far from complete, advised that the relationship between radiation dose and cancer induction was probably a linear (direct) one, and that it was not possible to exclude the chance that even low-dose radiation could cause injury to one or a few cells, which would then become cancerous. Because there were many gaps in the committee's knowledge, even this sensible position became the object of intense criticism by some physicists and medical scientists.

Nowhere did concern about the risks of low-dose radiation pivot more sharply than on mammography, a special X-ray examination of the breast.

Although mammography was first used in 1913, it wasn't until the 1960s, after the technique had been perfected, that it was widely employed. Breast cancer constitutes 27.2 percent of all cancers of women in the United States; each year approximately ninety thousand women are diagnosed as having it and thirty-four thousand women die of it. It is the most common cause of cancer in women, and before current treatments became available, only 18 percent of women with breast cancer survived for five years after being treated for it. Today, the five-year survival rate is about 60 percent, with a cornerstone of this improved survival being early diagnosis and treatment of the disease. The five-year survival rate can be raised to more than 80 percent if the cancer is treated before it spreads to other areas of the body.

Mammography became widely used in the 1960s, both for confirming the clinical diagnosis of breast cancer and for screening women to see whether they had it. What spurred the widespread use of this technique were studies indicating its specificity and sensitivity in diagnosing early breast cancer. The first of these was supported by the National Institutes of Health and undertaken by the M. D. Anderson Hospital in Houston, Texas. Mammography received a further boost in the mid-1960s when Shapiro, Strax, and Venet published the results of a four-year clinical trial. This study, often referred to as the HIP Study (Health Insurance Plan of Greater New York), concluded that repetitive screening, through the use of a clinical examination and mammography, led to a short-term reduction in breast cancer deaths.

These encouraging findings prompted the National Cancer Institute and the American Cancer Society to promote the Breast Cancer Diagnosis Demonstration Project (BCDDP), which began in 1973 and screened more than 250,000 women aged thirty-four to seventy-four in twenty-nine screening centers. The BCDDP had two major purposes: to detect breast cancers before they became palpable (obvious to the sense of touch) and to encourage physicians and women to use mammography as a routine screening technique. Participants in the BCDDP were taught breast self-examination and were given an initial physical examination, a mammogram (only for certain age groups), and a thermogram (a test that measures heat given off by the breast). The physical examination and mammogram were to be repeated annually for five years.

The BCDDP was by definition a demonstration project. Such projects are limited in time and scope, and aimed at testing out methodologies while simultaneously producing results. Because of design problems, but more importantly because of the controversy over the safety of mammography that erupted shortly after the project began, the project fell far short of its original goals, reaching only about one-fifth of the intended population. What results have been produced so far are often disputed and difficult to interpret.

The low-dose radiation used in mammography became the focus of controversy in the mid-1970s when Dr. John C. Bailar of the National Cancer Institute questioned the advisability of including women under the age of fifty in the BCDDP. He and others argued that the repeated radiation exposure of younger women—who had many years left before having to face the risk of breast cancer from any source—was causing more cancer than it was detecting. He and others also found that for the under-fifty age group, there were no data showing that the benefits of mammography outweighed the risk of the annual radiation exposure it involved. Bailar's criticism sent shock waves through the medical community, and startled women who had considered mammography the antidote to their breast cancer fears.

The National Cancer Institute and the American Cancer Society moved quickly to evaluate Bailar's assertions. In 1976 they appointed expert committees to assess the risks and benefits of mammography. One group, headed by Dr. Lester Breslow, then Dean of the School of Public Health at the University of California in Los Angeles, affirmed that there was no benefit from routine mammographic screening for women under the age of fifty, and substantial benefit in those over fifty. A second group, chaired by Dr. Arthur Upton, then Dean of Basic Sciences at the State University of New York Health Sciences Center in Stony Brook, concluded that there was a 1 percent increase in breast cancer risk with each 1 rad dose of radiation delivered to the breast during mammography. (The Upton group noted, however, that newer equipment was delivering only 0.1 rad during a routine mammogram.)

In order to study the findings of these groups, the National Institutes of Health convened a three-day conference on breast cancer screening in September 1977. Sixteen eminent scientists, epidemiologists, physicians, and representatives of the clergy, the legal profession, and the public reviewed all of the technical information and ethical issues surrounding breast cancer detection methods. On the basis of what the panel found, it recommended that mammographic screening be continued for women age fifty and over, but set limitations for those under fifty. For women aged forty to forty-nine who were enrolled in the BCDDP, the panel recommended mammography only for those who had personal histories of breast cancer or whose mothers or sisters had such histories. For women under forty, mammography was recommended only for those with a personal history of breast cancer. All of these recommendations were then incorporated into the BCDDP guidelines.

Despite the panel's recommendations, the controversy over mammography and breast cancer continues, hinging on the definition of risk. The American College of Radiology, going in a different direction, recommended an initial screening mammogram for asymptomatic women between thirty-five and forty years of age, followed by subsequent mammograms every one to three years. Dr. Stephen A. Feig of the Department of Radiology of the Thomas Jefferson University Hospital in Philadelphia pointed out in an article in the November 1979 *Journal of the American Medical Association* that the lower radiation doses delivered by newer equipment and the increased accuracy of newer techniques made the risk of mammography minimal, and mustered impressive scientific data to support this position. But others question it.

In July 1982 the American Cancer Society announced a change in its guidelines for using mammography. Citing the same arguments put forth by Dr. Feig four years before—the ACS recommended that the age for routine mammograms be reduced to forty rather than fifty years. The society's revised guidelines for breast cancer detection also recommended monthly breast self-examination beginning at age twenty, a physical examination by a physician every three years between the ages of twenty and forty and yearly after age forty, an initial mammogram between the ages of thirty-five and forty, annual or biennial mammograms thereafter from age forty through age forty-nine, and annual mammograms from age fifty on.

But disagreements continue, and the final word is not yet in on the risks of mammography, since no one has yet been able to precisely define its risks.

The controversy that has stalked mammography is now casting its shadow over several routine X-ray procedures.

Skull X rays: Officials of the U.S. Office of Technology Assessment told a Senate subcommittee in 1978 that too many skull X rays were being performed. They found that 20 percent were taken for trivial injuries and 34 percent as a safeguard against malpractice suits. The OTA concluded that routine skull X rays were of only limited benefit in the diagnosis and treatment of head injuries. Like many medical scientists, they pointed out that brain damage does not show up on such routine X rays, and that detecting such damage is really the crucial issue in treatment.

Chest X rays: In 1979, Dr. Herbert Abrams of the Harvard Medical School analyzed the reasons for the overuse of chest X rays. He cited poor techniques (requiring repeat examinations), patient demand, physician concern over malpractice suits, undue dependence on X rays in screening for tuberculosis, and institutional requirements. Most hospitals require that patients have a chest X ray when they are admitted, even if they have just had one elsewhere. Since patients aren't usually given their X-ray films—and are given a hard time when they ask for them— these previous X rays aren't readily available. Chest X rays are also routinely done before a patient undergoes surgery, even if there is no evidence of significant disease in a physical examination.

The doctors, patients, and hospitals simply want to play it safe. One study that reviewed preoperative chest X rays of 667 patients without chest symptoms found no abnormalities in patients under thirty years of age. In 38 percent of the patients a chest X ray had been taken during the previous year, and in 50 percent the only abnormality revealed by the X ray was an enlarged heart, usually detectable on routine physical examination.

A recent study by the FDA found that of the 75 million chest X rays done in 1980 (at a cost of $2 billion), roughly one-third were unnecessary because they were unlikely to reveal disease or alter its outcome.

Although the U.S. Public Health Service, the American College of Chest Physicians, and the American College of Radiology recommended in 1972 that chest X rays not be used in routine screening for tuberculosis, millions of people are still subjected to pre-employment chest X rays to check for this disease. Yet skin tests for tuberculosis are safer, more precise, and less expensive. Old habits die hard, and for some physicians there are considerable economic advantages in continuing to order (and charge heavily for) routine chest X rays.

Dental X rays: Every time a dentist takes a dental X ray, the part of the patient's skin closest to the machine receives about 1 rem of radiation, or ten times the average amount of background radiation that the entire body receives per year. Moreover, a 1979 FDA study found that one-third of dental X-ray machines (and almost half of breast X-ray machines) emit excessive radiation, although the dispersal of dental X rays is greatly reduced with machines that have a long, lead-lined cylinder instead of the typical short, plastic, pointed cone. Usually several X rays are taken at a time; and some dentists still insist on taking sets of X rays on their patients twice a year. If showered on the whole body, this much radiation would definitely be harmful; fortunately, it is less dangerous when concentrated on one area that is not especially sensitive (such as the mouth).

Nevertheless, the American Dental Association has advised, on the basis of new studies, that dental X rays not be taken routinely, but only when a dentist has no other way to get important information. Even for adults who get cavities frequently, X rays should be given no more than once every two years, and young children should not be X-rayed at all. And, says the ADA, all patients (particularly pregnant women) should be shielded during this procedure with rubber-covered lead aprons.

In 1979 Dr. Irwin D. J. Bross and two of his colleagues at the Roswell Park Memorial Institute in Buffalo, New York, caused a sensation when they published a startling paper in the *American Journal of Public Health*. Using data from a tri-state survey of leukemia cases in areas of New York, Maryland, and Minnesota, Bross and his associates showed that the incidence of leukemia increased as the radiation dose increased in ordinary diagnostic X rays; the 220 men he studied who had leukemia had been exposed only to the low radiation doses used in ordinary medical X rays. Bross estimated that the danger of low-level radiation causing cancer is tenfold greater than previously estimated.

So startling were these findings that the editor of the *Journal* printed a disclaimer along with the article, stating that "Dr. Bross stands virtually alone in defense of his data and the interpretations he places on them." The editor went on to say that the article was being published because of its controversial nature and because of its importance to public health. However, in the same issue of the *Journal*, the editor published a reply in the form of an article, written by Drs. John Boice and Charles Land of the National Cancer Institute, critiquing Bross' paper.

In their article, entitled "Adult Leukemia Following Diagnostic X Rays?", Boice and Land argued that the new statistical method Bross had used for analyzing the data from the tri-state study was invalid, leading to a gross overestimation of the risks of low-dose radiation. They characterized the new method as "too complex to be useful." In addition, they mustered historical evidence that did not support Bross' conclusions, including studies by other scientists that did not corroborate these conclusions and statistical biases and missing data that Boice and Land said had flawed the design of Bross' study. They agreed with Bross and his coworkers that no medical X ray should be performed needlessly, and that the doses of X rays should be reduced. They also said that there may be risks associated with every radiation exposure, and that no exposure should be assumed to be free of harmful effects. But, Boice and Land concluded, "society is not well served by exaggerating presumptive risks and excluding the possibility of benefit."

Thus, Boice and Land did not disagree with the Bross group

over whether or not there is a risk associated with medical X rays, but rather on defining the magnitude of that risk. The end result so far has been more controversy, which may in the end spawn the further research needed to precisely define the risks of low-level radiation.

The results of several recent studies have caused a number of noted radiation experts to estimate that low-level exposures should be judged anywhere from two to twenty times more dangerous than was previously thought. "I don't think there's cause for alarm," said Dr. Edward W. Webster of Massachusetts General Hospital (who suggests that risk levels may go up by a factor of four). "But the new estimates will put more heat into efforts to cut down radiation exposures, both medical and occupational."

Equally troubling is the disturbing evidence emerging on the potential dangers of nonionizing radiation. Unlike the ionizing variety, this type of radiation cannot knock electrons away from atoms and molecules in the human body. Because of this, it has long been regarded by scientists as much less hazardous than the ionizing kind—it was believed that it could not cause cancer—and so it has hardly been studied at all. But now this radiation, which springs from sources whose presence in everyday American life is almost ubiquitous—radio transmitters, power lines, radar, microwave equipment—is coming under increasing scrutiny. There has been a sharp rise in citizen protest and litigation concerning nonionizing radiation and some governmental efforts to curb exposures. Studies in Colorado and Sweden turned up evidence (not confirmed by other studies) that people living near electric transformers or power lines have increased leukemia or cancer rates.

Assessing these and other reports, Leonard Sagan of the Electric Power Research Institute (an industry trade group) said: "When you add it all up, it does appear that something is going on. But I don't think there is any reason to alarm the public." Lancet has noted that because nearly everyone is exposed to electrical and magnetic fields, "it is important to know what risks, if any, are entailed."

In June 1984 it was reported that the Environmental Protection Agency for the first time was recommending limits on the strength of radiation from radio and TV antennas because of possible human health risks. Recent studies had raised questions about whether "broadcast radiation" may cause disorders in the immune and nervous systems. If the EPA proposals were adopted, some broadcasters would be required to move their transmitters, raise their antennas, or cut their power—all of which could hurt reception in homes. But some radiation experts and community activists feel that EPA's proposed limits are still too high (and based on pure guesswork), and call for further study.

The results of one new study, reported in August 1984, were not encouraging. Researchers at the University of Washington, working on a project sponsored by the U.S. Air Force, said they had detected a higher rate of cancer among laboratory rats chronically exposed to low-intensity microwaves.

Another debate concerns risks from the low-dose radiation produced by defective CAT (computerized axial tomography) scanners, often called the "Cadillacs of medical technology." These machines—whose widespread introduction into hospitals in the United States during the 1970s revolutionized the practice of medicine—are basically combination computer-plus-X-ray machines that can take pictures of soft tissues such as muscles, which ordinary X rays can't do very well. However, in April 1983 the FDA announced that it had found an important defect in 238 CAT scanners designed and produced by the Technicare Corporation—which amounted to nearly 15 percent of the two thousand scanners now in use in this country.

The FDA's attention was drawn to problems with Technicare's scanners because of numerous complaints from hospitals and radiologists who owned them. The scanners tended to automatically repeat scans without the operator knowing it, thus exposing patients to excessive and hazardous levels of radiation. The FDA concluded that this tendency of the 238 scanners was due to a design defect rather than to wear and tear. FDA investigators also substantiated complaints that Technicare had never installed filters on sixty-four of the scanners, to reduce the amount of radiation called for in design specifications. This failure had led some owners of the machines to file civil suits. Prior to this FDA disclosure, Technicare had tried to persuade those owners who had demanded the installation of filters to buy newer Technicare CAT models instead.

The excessive radiation produced by Technicare's defective scanners was considerable, because they were used to perform some 775,000 scans annually. Yet F. Robert Niffin, a spokesman for Technicare, said that its scanners were "safe and effective," and counseled operators to continue using them. Under FDA pressure, the company installed filters on all of its machines that lacked them, and moved to correct the problem of repeat scans.

Concerns about the dangers of low-dose radiation associated with CAT scanners may soon dissipate because of the appearance of a newer technique known as nuclear magnetic resonance (NMR). This technique, which so far appears to be relatively safe, makes pictures of the body's tissues by exposing them to a magnetic field and measuring the responses of atomic nuclei in the tissues. These images are very similar to the cross-sectional X-ray pictures made by CAT scanners, but can be made without exposing patients to X rays. This, plus the fact that NMR produces sharper pictures and more clearly distinguishes various features in the body, gives it enormous advantages over CAT scans. NMR can also be used in place of currently widespread radiological procedures that use ordinary X rays and sometimes require the injection of dangerous chemical solutions to enhance the X-ray pictures.

Medical scientists expect NMR to become the predominant technique for diagnostic imaging at some point during the 1980s (it is now being used in many hospitals in the United States and Great Britain). One manufacturer of NMR scanners, the General Electric Company, estimates that the world market for these new machines will increase tenfold by 1988. Others foresee that NMR will quickly replace CAT scanners in most large hospitals and medical centers. But the cost of the new machines, an average of $3 million per unit, will preclude their purchase by smaller community hospitals for some time.

It is entirely possible that because of newer technologies such as NMR, the problem of low-dose radiation associated with medical and dental X rays may eventually become a historical curiosity. For now and the foreseeable future, however, X rays will still be widely used, and low-dose radiation from a plethora of other sources, natural and manmade, will be a continuing concern. The question "How much is too little?" still has not been answered.

5
Government
Guardians

Government regulations may not make life fair or force the trains to run on time, but they do—and this is no small thing—reduce chaos and anxiety. Even in an age when large segments of the public routinely denounce "big government," and industry officials charge that regulations are strangling business, each of us, almost every hour, receives some regulatory protection for which we are grateful.

Americans may be individualists, but they are far from being anarchists; few oppose all regulation. Nevertheless, so many people are against so many regulations that it sometimes seems as if we should abolish the dozens of regulatory agencies and thousands of statutes that have come into being in the past two dozen years and start all over again. Some have even gone so far as to suggest that the business of government is no longer business but regulation. Ronald Reagan liked it better the old way, so when he became President he tried to halt the torrent of federal rules. The least that can be said for this effort is that it sparked a welcome political debate.

The popular perception of deregulation today as a conservative, pro-business cause belies its historical context. "It is important to recall that virtually the entire regulatory system was built in response to the needs of the business community," Susan and Martin Tolchin pointed out in their recent book *Dismantling*

America. Even now owners of steel plants may oppose certain environmental regulations while strongly supporting regulations that restrict steel imports. Liberals such as Senator Edward Kennedy back deregulation of the trucking and airlines industries, while conservatives such as Jesse Helms stoutly defend regulations that maintain tobacco price supports.

Most regulations are not "good" or "bad," although *some* regulations are. People tend to seek out new rules when they serve their purposes, and oppose them (usually on financial grounds) when they do not. The Tylenol poisonings in 1982 nearly drove the pharmaceutical industry into a panic—until they sought, and received, new federal regulations requiring tamper-proof tops that reassured consumers and maintained sales.

Yet it would be wrong not to recognize that the debate over regulations in the health and safety field has certainly been fired by ideology. It began innocently enough with President John F. Kennedy's call for justice and product safety in his 1963 bill of consumer rights. Shortly afterward, in rather rapid order (considering the generally slow pace of change in Washington), a plethora of new agencies and administrations were formed, adding to the already rich federal alphabet soup; these included the EPA, CPSC, OSHA, NHTSA, NIOSH, to name just a few. A new word—consumerism—became popular, and a new media celebrity (Ralph Nader) was born. During the 1970s, the number of federal regulatory agencies nearly doubled, the full-time personnel at these agencies (and the number of pages of regulations in the Federal Register) almost tripled, and the agencies' total annual budget increased sixfold.

But as Ralph Nader later commented, the new agencies and laws worked so well that they spurred a corporate counterattack. Between 1974 and 1980, the United States Chamber of Commerce doubled its membership and trebled its annual budget. High-level executives formed the Business Roundtable and hundreds of individual trade associations sprang up to lobby Congress. A sure sign that consumerism was in deep trouble appeared in 1977 when Congress dropped plans to establish a federal consumer advocacy agency. With the election of Ronald Reagan, the consumer movement—as it was often said—had gone from Nader to nadir.

In a speech during the 1980 campaign, Reagan had said that "free enterprise is becoming far less free in the name of something called consumerism." Following his inauguration, he announced that "the major problem concerning the consumer today is too much government," and set up a Task Force on Regulatory Relief. An official administration policy statement said that "While regulation is necessary to protect such vital areas as food, health and safety, too much unnecessary regulation simply adds to the costs to businesses and consumers alike without commensurate benefits." The Reagan Administration set out to reduce "paperwork" with the same fervor environmentalists had sought to reduce pollution.

To combat the proliferation of regulatory weeds slowing America's growth, Reagan ordered that all proposed and final government regulations be submitted to his budget office for review. And the President appointed administrators who were sympathetic to his get-the-government-off-the-backs-of-the-people approach.

"Hogs are like people," said the new Secretary of Agriculture John R. Block. "You can provide protein and grain to a hog and he will balance his ration. People are surely as smart as hogs. I am not sure this government needs to get so deeply into telling people what they should or should not eat."

The results of the Reagan deregulation drive were impressive. During Reagan's first two years in office, the federal government proposed about 25 percent fewer regulations than it had during the same period under Reagan's predecessor, Jimmy Carter. Cutting agency budgets practically insured that the agencies would issue even fewer regulations in the years ahead. From 1980 to 1983 the EPA's budget dipped 30 percent, the National Highway Traffic Administration's budget 37 percent. The number of personnel at the Occupational Safety and Health Administration fell by 23 percent.

These cutbacks not only stemmed the flood of regulation coming from these agencies, it also severely hampered their enforcement capability. To many it appeared that in those instances in which the Reagan Administration could not stop regulations from going into effect, or rescind them, it did the next best thing: *de facto* deregulation through lax enforcement. Conscience is often nothing more than acting as if someone might be watching, H. L. Mencken suggested, and now the watchdogs in the federal government were becoming increasingly scarce. Inspections by the OSHA fell off by 17 percent during Reagan's first year in office; enforcement actions at the Food and Drug Administration were halved; and the EPA referred only sixty cases to the Justice Department for possible prosecution, as compared to the customary annual total of two hundred. "If we don't enforce the laws on the books," warned Jeffrey Miller, acting administrator for enforcement at the EPA during the Carter Administration, "there is no incentive for companies to comply with them."

Although it did not receive the media and Congressional attention lavished on the EPA (which had become a symbolic battleground) during this period, the Consumer Products Safety Commission suffered effects from the Reagan cutbacks that were perhaps equally severe. (Some twenty-five thousand people die and 34 million are injured each year in accidents involving consumer products.) The 25 percent cutback in funds for the CPSC, just one innocuous line in the overall federal budget, translated into the very real effects of a personnel cutback by one-third, a 47 percent reduction in field investigators, 50 percent fewer inspections, the closing of three test labs, and ten regional offices reduced to five. Perhaps because of a perception of less rigorous CPSC enforcement, reports of faulty products volunteered by manufacturers fell from 147 in 1980 to 96 in 1982. "We cannot afford any further budget cuts," CPSC commissioner Nancy Steorts protested. "This agency has taken as much as it can handle." Despite this, the Reagan Administration called for another 6 percent CPSC budget cut for 1984.

The President gained plenty of support for these policies in Congress, which read his landslide victory in 1980 as a mandate for regulatory relief. The CPSC discovered this in 1981 when its new safety design standard for power lawn mowers, of which it was quite proud, was overruled by Congress, where many contended that the rule would raise consumer prices excessively. As a result, "the same eighty thousand people who are injured each year by the cutting blade will still be at risk," CPSC commissioner R. David Pittle noted.

But the word for these actions (and inactions) was spelled r-e-l-i-e-f, and according to the Administration, regulatory changes were saving business about $2 billion in annual operating costs and about $5 billion in one-time capital investment costs. (It was too early to tell whether any of these savings would indeed be passed along to consumers.) Vice President George Bush, head of the Task Force on Regulatory Relief, announced in August 1983 that the group was going out of business because "the executive branch has done what it can"; the next steps were up to Congress. The Task Force said the Administration's actions would save consumers and business more than $150 billion over the next decade, and reduce by 300 million hours the amount of paperwork that businesses were required to do. Consumer advocates, however, challenged these figures and accused the Reagan Administration of turning back the clock on progress in cleaning up the environment and improving health and safety.

The Republicans nevertheless continued to claim that a new era had dawned in America—or was it the return of an old era whose passing had been largely unlamented? But the Administration wasn't through. It said that it preferred "market alternatives" to costly federal regulation. This was code language. The 1980 Republican platform stated it more clearly: "An informed consumer making economic decisions in the marketplace is the best regulator of the free enterprise system."

This, of course, assumed that the vast majority of consumers are informed. Indeed, the now chairman of the Federal Trade Commission, James C. Miller III, after announcing that "consumers are not as gullible as most regulators think they are," proposed that the FTC stop enforcing truth-in-advertising rules. Virginia H. Knauer, the White House's consumer spokeswoman, proposed putting manufacturers on the honor system, saying that "consumers are very sophisticated, and by making their own choices in the free market they become the regulators."

Miller's call for huge FTC budget cuts and a regulatory slow-down led *The New York Times* to note: "Some zealots of yore at the agency are more restful now, and the most popular corridor expression used [at the agency] lately to describe the attitude of prescribed languor, according to one commission source, is: 'Now comes Miller time.' " Yet the FTC still became, under Miller, quite controversial. Representative Albert Gore of Tennessee charged that "some screwball" at the FTC had suggested in a memorandum that market forces, in the form of wrongful death lawsuits by surviving spouses, might be more effective than an agency recall of six thousand defective survival suits used by seamen. Miller and his Bureau of Consumer Protection chief, Timothy J. Muris, caused another storm when they called for rescinding certain rules to ensure that advertisers could actually substantiate claims they made in their commercials. "When the television weather reporter states that 'it will rain tomorrow,' the typical viewer does not interpret the statement as meaning that rain is absolutely certain," Muris wrote in one memo. "The same analysis applies to advertising claims."

One problem with this approach, said Rhoda H. Karpatkin, executive director of the Consumers Union, was that it would move the country "back into the age of 'Let the buyer beware,' or maybe even, 'Let the buyer be milked.' " The other problem was that the poor could not afford it. FTC chairman Miller admitted that "avoiding defects is not costless. . . . Those who have [a] low aversion to risk—relative to money—will be most likely to purchase cheap, unreliable products." In a policy statement adopted by a three-to-two vote in October 1983 the FTC indicated that it would no longer take action against advertisements that seem deceptive on their face. Instead, the commission, before filing a complaint against the advertiser, would have to first prove that a "reasonable" consumer had suffered actual injury.

Most research indicates that consumers are not, at heart, rational purchasers. They often buy on impulse and without lengthy debate over the costs and benefits of a product. Perhaps that is why—despite all the antigovernment rhetoric—surveys show that most consumers back strong food and product safety laws, truth-in-advertising rules, and tough environmental regulations at whatever cost.

In their landmark study in the mid-1960s, pollster Lloyd Free and psychologist Hadley Cantril concluded that Americans are at once ideologically conservative and operationally liberal. Most

Americans, they reported, "continue to accept the traditional American ideology which advocates the curbing of federal power," while "favoring government programs to accomplish social objectives." The majority, they found, agreed with general statements about the federal government interfering too much in local affairs and moving toward "socialism," but were very supportive of specific antipoverty and health care programs. "In typically pragmatic American fashion," they wrote, 'the practical is given precedence over the theoretical."

Nearly two decades later, Free declared that he had little doubt that the same "schizophrenia" still existed. "It's quite clear," he said, that "Americans haven't changed in terms of what they feel the government ought to be doing. . . . But if the old conservative philosophy weren't still relevant, Ronald Reagan would never have been elected in 1980. Americans love to hear him talk about the private enterprise system and they love the lines he takes in terms of antigovernment. It rings these ideological bells."

Recent survey results dramatically confirm these observations A 1980 Yankelovich Poll showed "a dramatic rise in antigovernment sentiment. . . . Strong majorities of Americans viewed the government as inefficient and bogged down in red tape; most Americans believed the government spends too much money on wasteful programs." Yet the polling firm advised its forty corporate clients that "antigovernment sentiment is really an unhappiness with cost and inefficiency, rather than reflecting a belief that government should not play a major activist role in society."

A 1981 Roper Poll showed that 55 percent of those surveyed felt the federal government had gained disproportionate power at the expense of local and state government, but a substantial majority said they wanted the federal government to have a major role in protecting the environment and guaranteeing good health care. At the same time, a Harris survey was showing that while 59 percent believed that "the best government is the government that governs least," 78 percent agreed that Washington "should regulate major companies, industries and institutions to make sure they don't take advantage of the public."

This interventionist sentiment—so counter to what most observers (including President Reagan and his top consumer and regulatory aides) contended was the prevailing public mood—is even more explicit when the issues become more specific. A recent study by Seymour Martin Lipset and William Schneider of the American Enterprise Institute found that only 8 percent of the public favor less regulation in the truth-in-advertising area, and just 11 percent favor less on product safety. Moreover, the public feels by 52 to 12 percent that the costs of regulations to protect workers' health and safety are worth it, and 42 versus 19 percent feel the same way about environmental laws.

Perhaps most significantly, an October 1982 Harris Poll survey, entitled "Consumerism in the '80s," led the pollster to declare that "there is every sign of yet another explosion of consumer concern in the marketplace from one end of this country to the other." Congress got a 71 percent negative rating for protecting consumer interests; President Reagan's negative rating was 65. In questions about nine consumer problems, survey respondents indicated that seven had become *more severe* since 1976. And by a large majority the public requested greater health and safety regulation.

Opinion polls that don't force individuals to make "trade-offs" are not very revealing, risk assessment expert William W. Lowrance charges. And perceptions, he says, are subject to rapid shifts. "Hydropower will probably continue to be viewed as harmless," Lowrance points out, offering one projection, "until the day a large dam breaks over a major population center." But there can be little question that while choosing to stand their ground ideologically, many Americans are in fact running scared when it comes to their own well-being, and would like a little reassurance from Uncle Sam. As Lloyd Free once commented, what Americans want is not an *end* to "big government" but *better* big government. We have grown used to—if not comfortable with—government guardianship, and it may be too late to turn the clock back as far as conservatives would like to turn it. Individuals will actually need more, not less, protection as technological and scientific advances—such as genetic engineering—take us all into unknown realms with unprobed dangers.

It has been said that the government regulates only when

public pressure builds to make it regulate. After a few more years of dealing with some of the negative effects of deregulation, that pressure—ideologically impure as it may be—may start building again.

Close-Up (1)
THE BATTLE OVER
THE AIR BAG

"It's too easy a solution to say, 'Make cars safer.' Believe me, if we could do it, we would."

—Roger Maugh, Safety Director,
Ford Motor Company

On April 27, 1971, President Richard M. Nixon met with Henry Ford II and Lee Iacocca, then the two top executives of the Ford Motor Company, in the Oval Office at the White House.

Although rumors about this meeting circulated in Detroit and Washington for more than a decade, details were not verified until 1982, when a transcript of a White House tape recording of the meeting surfaced during a wrongful-death suit filed against Ford by the parents of a teenage girl killed in a crash of one of the company's cars.

The transcript showed that Ford and Iacocca had wanted Nixon to rescind a pending federal regulation requiring air bags (or another "passive restraint" system) in every new car. The two auto executives acknowledged that new safety standards were saving lives, but argued that Ford desperately needed regulatory relief. The Department of Transportation, Iacocca asserted, was "not willfully but maybe unknowingly . . . really getting to us" by insisting that "the citizens of the United States must be protected from their own idiocy." Iacocca said that the cost of safety features in Ford cars would rise by $250 with a mandatory air bag rule. He noted that foreign car makers would have to install air bags too, but complained that they could do so at lower cost because they paid their workers less.

"We are in a downhill slide, the likes of which we have never seen in our business," Iacocca told Nixon. "And the Japs are in the wings ready to eat us up alive. So I'm in a position to be saying [to the Department of Transportation]: 'Would you guys cool it a bit? You're gonna break us.' "

Henry Ford pointed out to Nixon that "There are many things in DOT . . . that you could do [by] just callin' 'em up. I'd just say, 'Well, let's get some cost-effectiveness.' "

Nixon agreed that "cost-effectiveness is the word." He sympathized with the Ford executives. "We can't have a completely safe society of safe highways or safe cars and pollution-free and so forth," said Nixon. This, he said, would require people to go back to living like a "bunch of damned animals. . . . The great life is to have it like when the Indians were here. You know how the Indians lived? Dirty, filthy, horrible." Nixon dismissed environmentalists, consumer advocates, and auto safety activists as "a group of people that aren't one really damn bit interested in safety or clean air. What they're interested in is destroying the system. . . . The safety thing is the kick 'cause Nader's running around squealing about this and that and the other thing."

Impressed with the Ford executives' argument, Nixon said he would look into the matter, and told them that his domestic affairs adviser, John Ehrlichman, would be their "contact" at the White House. Although it is impossible to prove cause-and-effect in this matter, the record does show that Ehrlichman, in a confidential memorandum, subsequently ordered the Department of Transportation to delay imposition of the air bag regulations.

Following the release of the transcript in 1982, Ben Kelley, senior vice president of the Insurance Institute for Highway Safety, said the Nixon/Ford/Iacocca meeting "and what apparently flowed from it have made a very important contribution to the fact that, still today, Americans cannot buy air bag protection at any cost in any car."

Although the White House conference between the President and the auto executives was extraordinary, it revealed, in the most pedestrian manner, the gut issues on which the fifteen-year battle over the air bag has been fought: economics, ideology, and emotion.

When in 1968 the National Highway Traffic Safety Administration (NHTSA) made the installation of lap and shoulder belts
mandatory in all cars, officials of that agency were well aware
that these "active" restraints would be of little use to the majority of Americans who would not activate them. Thinking it
knew what was best for today's motorists, the NHTSA began
testing "passive" restraints, such as the automatic seat belt and
the air bag (which, from its position under the dashboard, inflates instantaneously on impact, protecting passengers from
smashing into hard metal or glass).

When a car crashes into a wall at forty miles per hour, for
example, its front end is crushed and the vehicle slows down
drastically. But the person inside the car has nothing to slow
him or her down, and so continues to move forward at 40 mph.
Within 1/10 of a second the car has come to a complete stop
but the motorist is still moving forward at 40 mph. Within one-
fiftieth of a second after the car has stopped, the occupant slams
into the dashboard, windshield, or both. Ninety-two percent of
persons killed in car crashes are front seat riders.

It is within this crucial one-fiftieth of a second that restraints
come into play. Seat belts across the lap and chest hold the
motorist in place; an air bag inflates within one-fiftieth of a
second to cushion the motorist's forward motion.

In 1971 the NHTSA amended its occupant crash-protection
standard No. 208 to require passive restraints in all cars beginning two years later. The nation's largest auto maker, General
Motors, seemed enthusiastic about the idea, perhaps because
it had the jump on its competitors in air bag technology. But
following the Nixon/Ford/Iacocca meeting, the air bag requirement was postponed, pending further study, in favor of the
"ignition interlock" system, which (as described in Chapter 1)
may have been a better idea for the Ford Company but was a
disaster for the public.

Despite this postponement, General Motors proceeded with
production of the air bag, offering this safety device as an option
on its automobiles from 1974 to 1976. It sold only ten thousand
air bags.

One problem was the price. Mass-produced, the air bag would

have added little more than $100 to the cost of a car ($200 or more today); in limited production the price soared to $415. A second problem was that Americans doubted the reliability of air bags, many fearing that these devices would inflate if their car even tapped the car in front of them while parking, or suddenly inflate for no apparent reason while they were speeding along an interstate highway.

Studies showed that air bags rarely malfunctioned—the DOT computed a figure of one inadvertent inflation per 3.3 billion vehicle miles driven. In one test for General Motors forty drivers were suddenly surprised by air bag deployment, and not one lost control of his car (the bags quickly deflate). Several studies, meanwhile, calculated that if universally installed, air bags could cut traffic fatalities by about one-fifth each year. Still, public comments received by the DOT ran four to three against requiring the air bag as standard equipment.

Although it could have outflanked Ford and most of its foreign competitors (who were not then air-bag-capable) with a strong marketing pitch for these safety devices, GM refused to do so. Thus, while GM's engineers won plaudits in the scientific community for their air bag work, and GM won an industry award for its testing program, the company's press releases and official pronouncements on the air bag were pessimistic, and GM did not pressure its sales force to make a big pitch for this option.

Why? Some have attributed this behavior to pique on the part of GM executives, who were angry at having spent millions of dollars for air bag research (on behalf of bullheaded drivers who would not wear their seat belts) only to have Ford outlobby them in Washington. GM President Edward Cole later admitted that the company did not "create a desire on the part of the user" to purchase the air bag. Many dealers, according to surveys, openly discouraged customers from buying them.

Dr. Arnold Arms, a Kansas City physician, survived a head-on collision with a bus in 1975 because he was driving an Oldsmobile equipped with an air bag. But when he tried to buy a new Olds with an air bag, the dealer put him off for so long that Dr. Arms never got another car with the safety device— GM stopped offering the option in 1976. A GM spokesman told

The Wall Street Journal that most car buyers simply didn't like air bags, and that Dr. Arms was "biased" in favor of the device. "Anybody whose life has been saved by the air bag would be biased in favor of it," said the GM official.

Yet a survey conducted for GM in 1975 showed that one-third of all Oldsmobile buyers would have purchased a car with an air bag if it had cost them only an extra $100 to do so. In 1976 this reasonable, attainable goal (based on mass production) seemed a long way off.

The mood changed, however, when the Carter Administration took command. On June 30, 1977, the new Secretary of Transportation, Brock Adams, noting the projected rise in the auto death toll caused by the spurt in small-car sales, mandated that all cars manufactured after September 1, 1983, be equipped with some form of passive restraint. Adams declared that passive restraints would save twelve thousand lives a year and could be installed at the cost of $112 (for air bags) or $25 (for automatic seat belts). Public doubts about the devices should not be taken into account, Adams said.

The auto industry was not happy with this requirement. Chrysler distributed an editorial cartoon to newspapers showing an air bag spilling out of a car window with a gas station attendant informing the driver, who is struggling to extricate himself, "Honest, all I did was slam the hood." On the other hand, auto safety activists felt that the DOT timetable was too extended. Still, it appeared that the air bag was back on line.

This changed again when another new Administration came to Washington. During the 1980 Presidential campaign Ronald Reagan had told a Detroit audience that "federal regulations are the cause of all your problems," and once in office he did something about it. For starters, he cut the NHTSA budget nearly in half.

On October 23, 1981, the DOT rescinded the latest passive-restraint regulation, reportedly overruling the advice of top staff members of NHTSA. Raymond A. Peck, Jr., NHTSA chairman, explained that most manufacturers were planning to comply with the measure by installing automatic belts, not air bags. Peck said that while air bags were more effective than ordinary seat belts, automatic belts were not. Because motorists feared entrapment after a crash, the NHTSA had required that automatic belts be readily detachable. Peck stated that automatic belts were, therefore, fatally flawed because, with a little tinkering, the devices could be permanently deactivated by motorists.

Why, then, didn't Peck simply eliminate the belt option and make air bags mandatory? Peck explained that he was "extremely hesitant" to do so, partly because "quantum leaps forward in technology bring industry lawsuits," and the possibility of air bag failures might lead to liability claims. (Peck was being somewhat disingenuous, since he had to know that air bags, in tests, had failed infrequently.)

But most important, according to Peck, was that installing air bags could cost the American auto industry $1 billion a year, and that from a cost-benefit perspective (which the Reagan Administration urged), this $1 billion, divided by the minimum of 9,000 lives that would be saved, meant that the auto industry would spend about $110,000 to save each human life.

The Administration, then, had decided that a human life was not worth $110,000—an honest opinion but one open to debate. Indeed, by the DOT's own estimate, made in 1975 and updated to 1981 dollars, each auto fatality costs society $480,000.

"Peck," argued Colorado Congressman Timothy Wirth, "has today signed the death warrant for thousands of Americans, while sentencing millions more to a life of serious disability and injury. . . . He has not only abrogated his responsibility as the person charged with keeping American motorists safe, but he has cost the public billions of dollars in payments for support of the families of the dead, and for the care of those who are seriously disabled by automobile accidents. The tragedy is that the technology exists to save these lives cheaply, but the auto industry refuses to use it."

Wirth went on to point out that "In passing the National Traffic and Motor Vehicle Safety Act, Congress did not intend that safety requirements would move up and down with the fortunes of the general economy or with the economic conditions of the automobile industry in particular." Even if GM, for example, was saving half a million dollars daily because of the air bag delay, it still continued to post millions of dollars in losses nearly every quarter. Roger B. Smith, GM chairman, looked on the bright side. "If nothing else," he said, referring to the halt in regulations, "our hearts are lighter."

Critics of the Administration's move pointed to figures showing that air bag installation would *save* money—$2 billion a year in auto insurance costs and $2.5 billion in medical expenses and health insurance costs, according to some estimates—while producing needed jobs making and installing the new devices in Detroit. The Allstate Insurance Company calculated that for policy holders who drove a car with an air bag, the savings on insurance premiums alone would pay for the option within four years. Critics also charged that the auto makers saw safety regulations as an irritant which they had had to tolerate for far too long, and had set out to dismantle the passive restraint rule.

By this time, the air bag issue had become more than just another regulatory nightmare. It was turning into a symbolic, ideological battle.

Insurance companies and consumer groups petitioned the U.S. Court of Appeals for the District of Columbia to set aside Peck's decision, and on June 1, 1982, the court temporarily blocked the repeal of the passive restraint rule. Using unusually harsh language, Judge Abner Mikva wrote that the NHTSA had acted "arbitrarily and capriciously" in revoking the regulation. He said that "it is difficult to avoid the conclusion that [the decision was] distorted by solicitude for the economically depressed automobile industry—which is not the agency's mandate—at the expense of consideration of traffic safety, which is. . . . Essentially, the agency seems to conclude that because some technology won't meet the 'passive-restraint' standard, it needn't mandate compliance by technology that will. The absurdity of this Orwellian reasoning is obvious."

Mikva further declared the NHTSA's analysis of air bags to be "nonexistent," stating that the agency had "failed or flatly refused to evaluate the cost-effectiveness of these devices."

NHTSA is correct in expressing concern about negative public reaction to "an expensive example of ineffective regulation." There is no basis on the record before us, however, for concluding that Modified Standard 208 is such a regulation. More important, it is erroneous to believe that "ineffective regulation" occurs only when government acts affirmatively. By rescinding the passive restraint standard without legal justification, NHTSA's arbitrary action presents a paradigm of ineffective regulations. (It) has wasted administrative and judicial resources . . . The agency's action here thus represents "an expensive example of ineffective regulation" of the worst kind.

Following this ruling a Ford spokesman said the company still believed that using existing seat belts, for which motorists "have already invested more than $14 billion," was still "the most effective strategy" for safety. On September 9, 1982, the Justice Department announced that it would take the case to the Supreme Court.

Nine months later, the Court, charging the Reagan Administration with being "arbitrary and capricious," ruled unanimously that the automatic crash standard should be reinstated. In his written opinion, Justice Byron White observed that the auto industry had "waged the regulatory equivalent of war against the air bag." The NHTSA, he said, "may not revoke a safety standard which can be satisfied by current technology simply because the industry has opted for an ineffective seat belt design." But the decision left the NHTSA free to revoke the safety regulation once again if it could come up with a better justification for doing so, and there was speculation that the agency would indeed do just that.

While the Reagan Administration was trying to figure out how to puncture the air bag solution, auto makers overseas were beginning to embrace it.

In March 1981, Mercedes-Benz, in full-page advertisements in West Germany, called the air bag "a good idea whose time has come." The company, continued the ad, had developed "the first air bag safety system in the world ready for series production." Three years later, the air bag was offered in America as an option costing $880 on certain 1984 models, but Mercedes hoped it would eventually become a low-cost standard feature. (Perhaps because of past marketing problems, Mercedes-Benz now referred to the device not as an air bag but as the "Supplemental Restraint System.") For the time being, their marketing program would be considered a success, said Mercedes officials, if air bags were ordered in 10 percent of their cars. BMW and Volvo may offer air bag options soon.

Meanwhile, the *National Journal* in Washington, D.C., reported rumors that Japanese auto makers were "seriously considering the possibility of equipping their American-based automobiles with air bags regardless of whether the U.S. government mandate[d] them," and added that if the Japanese correctly concluded that there was a potential American market for air bags, "the impact could be shattering on Detroit . . . [approaching] what happened when Japanese and other foreign manufacturers jumped ahead of the Americans with small, fuel-efficient, low-priced cars."

So there would seem to be competitive, as well as safety, reasons for accepting air bags; the lives American auto makers save might be their own. Why has this taken so long to transpire?

Public hostility to the air bag seems to be fading. A Gallup Poll found in July 1984 that Americans now favor installation of air bags in all cars by 60 to 31 percent, with young adults voting overwhelmingly in favor. There seems to be a "gender gap" emerging—women back the bag by a much higher margin than men.

However, one survey revealed a major reason why the popular will has been stymied for so long. A *New York Times* national poll in 1980, which showed the general public in favor of a mandatory air bag regulation by 45 to 32 percent, revealed that 93 percent of retail auto dealers *opposed* the idea.

In fact, virtually every survey has indicated that the public is less disturbed by the air bag than the prospect of a new, national law that would require drivers and passengers to take action, in the form of buckling their seat belts. This indicates that the public believes that it is sometimes the government's duty to take drastic action (such as installing air bags) to protect people from their own laziness.

"Modern societies long ago learned that, whenever practical, the best ways to reduce damage to people from hazards in the environment are those that do not require the people to be always expert, alert and to take evasive action," William Haddon, Jr., head of the Insurance Institute for Highway Safety, has written. "Thus we insulate household wires, rather than telling people never to touch them; we put fuses in electrical systems, rather than saying that somehow we should always manage to prevent short circuits; and we purify water supplies and pasteurize milk, rather than telling people they should always boil them to prevent illness. This correspondingly removes from people the necessity to be perfect, as well as the blame when they are not."

This, of course, is not a new view. It was, for example, stated in Deuteronomy this way: "When you build a new house, you shall make a parapet for your roof, that you may not bring the guilt of blood upon your house, if anyone fall from it."

The automobile industry has never accepted this view, citing the unbearable costs of safety devices. For the same reason motorists have accepted the fact that, in regards to safety, the cars they are driving are less than state-of-the-art.

For its part, the Insurance Institute believes that automobile makers deliberately overstate the cost of safety factors, and that their failure to implement further safety improvements is indefensible. In fact, it believes that in terms of crash protection, today's cars "are literally obsolete." It contends that for years it has been both technologically feasible and economically practical to design, build, and equip cars to make them far more protective of their human occupants, and that for every year this potential goes unfulfilled, hundreds of thousands of people suffer needlessly.

One example of this is the twenty thousand Americans who develop epilepsy each year as a result of auto accidents. Addressing this, Tony Coelho, a California Congressman who has epilepsy produced by a head injury incurred in an auto accident, points out that "It costs far more to deal with epilepsy, not only in terms of dollars but also in terms of human frustration and in terms of lost productivity, than it could ever cost us to continue our efforts to reduce auto-accident-related injuries which cause epilepsy."

But still, Detroit and Washington have moved fitfully on the air bag question. Why?

In an editorial *The New York Times* observed that "Detroit lives by axioms," one of which is "Safety Doesn't Sell." The newspaper recalled that many years earlier, the Ford Motor Company had tried to sell safety, "flogging the virtues of padded dashes in its ads and waiting for customers to flood in," but that "people seemed to prefer Chevys and Dodges with bathing beauties on the hoods and fins on the tails."

The "Safety Doesn't Sell" axiom, however, may be not only self-serving but outdated. Most surveys show that prospective car buyers rank safety features *behind* cost, gas mileage, and repair record as considerations in making their choices, but *ahead* of comfort, styling, and dealer service.

Auto makers argue that the government should let them voluntarily develop new safety features for their cars and freely offer these to the public. Safety decisions should be determined in the marketplace, they say, with new car buyers simply picking and choosing among various models and deciding how much certain safety options are worth to them.

But former NHTSA administrator Joan Claybrook argues against this approach, observing that "If safety technology should be incorporated into cars, the public is not served by having it adopted in some and not others. This is confusing, and it is difficult for car buyers and users to keep track of which makes and models have safety improvements. More importantly, the sanctity of life for buyers of Hondas is no greater than it is for buyers of Chevrolets. . . ."

Another argument offered by officials who loathe the air bag is that it is patently unfair to charge people who dutifully wear their seat belts for the cost of installing a somewhat unnecessary (for them) air bag in their cars. In fact, however, a 1978 survey done for the NHTSA showed that support for air bags was *strongest* among those who regularly wore seat belts.

"The air bag," says the Insurance Institute's Ben Kelley, "is so closely and symbolically tied to the notion of government regulation that Detroit can't give in because it would show that it lied about overregulation for years." Thus, says Kelley, Detroit is "trapped in its whole history. So trapped that it runs the risk of getting eaten alive on air bags by the Germans and Japanese. It's ironic, because the auto industry, like most industries, *wanted* regulation at the start, to provide stability and protection. They want basic federal regulation, otherwise they'll get killed by having to adapt to different state standards. But later, like we saw with the airlines, they want less regulation, or even deregulation."

What has kept the air bag alive for so long, Kelley says, is that it never became a captive of the typical liberal/conservative conflict over government intervention. Many conservatives, such as George Will, have supported the air bag, "because they can't understand why they don't at least have a *choice* in the matter," Kelley observes. "They wonder why it's not available, at least, as an option." (Raymond Peck himself expressed interest in buying an air bag for his car.) They realize, in addition, that when an individual is injured unnecessarily *everyone* pays; roughly half of the medical costs of auto accidents are borne by people far from the scene of the crash, in the form of government assistance (provided by taxpayers) and higher insurance premiums.

"You can only stop technology for so long," Kelley says. "It's like with the air brake on trains. The railroads always blamed accidents on the 'drunk Irish brakeman.' It was a convenient excuse. It took thirty years to get air brakes installed after they were developed. It was inevitable. It's just a shame that thousands suffered, waiting."

Despite favorable court rulings it was by no means "inevitable" that the air bag would join the air brake as standard safety equipment. Four months after the Supreme Court handed the ball back to the Reagan Administration, the President's new Secretary of Transportation, Elizabeth Dole, punted. Secretary Dole proposed neither a mandatory air bag rule nor a new standard for automatic seat belts. Instead she called for a six-month review, to gather and analyze new data on costs, benefits, and acceptability to consumers.

"I have no mandate higher than safety," Dole said. "We're not dragging our feet or slowing down the process." But she admitted that the latest delay would mean that automatic crash protection equipment could not appear in cars until the 1987 model year, at the earliest. In the meantime, critics pointed out, another thirty thousand Americans would die needlessly. Paradoxically, just days before announcing this decision, Dole proposed an important regulation that called on all cars produced after September 1, 1985, to have a third brake light mounted at the rear of the car at drivers' eye level. Like the air bag, the rear light would add to the cost of a new car but would not eliminate nearly as many injuries per year (forty thousand vs. over one hundred thousand for the air bag).

Finally, on July 11, 1984, Secretary Dole announced her decision: All cars sold after September 1, 1989, would have to be equipped with passive restraints. But the restraining devices, she said, could be air bags or automatic seat belts, and the requirement would be waived entirely if, by April 1, 1989, states representing two-thirds of the population of the United States enact mandatory seat-belt laws.

Outside Detroit the reaction was not favorable. The State Farm Insurance Company filed a lawsuit charging that Congress had vested the authority to make this rule in the Department of Transportation, not state legislatures. "It's a snare and a delusion," Ralph Nader said. "It destroys any prospect of air bags being installed." Bill Haddon pointed out that in Ontario only about 60 percent of the drivers buckle up, despite a mandatory law, and that this figure is much lower at night and among those likely to be involved in crashes.

Many auto safety activists were perplexed by Secretary Dole's lack of commitment to the air bag. She has had an air bag in her own car ever since she was involved in a severe accident, caused by a drunk driver, in the mid-1970s, and still suffers from back pain caused by the crash.

It was beginning to look as if the best hope for the arrival of the air bag lay not in regulation but in America's penchant for litigation. When the U.S. District Court in Washington awarded damages against Fiat, the Italian automaker, for failing to provide state-of-the-art safety design for its door locks, one veteran auto industry attorney commented: "The company met the federal standards for door locks but the court said that it failed to use the best system that was available. That's the way it is with the air bag. We know what it can do but we don't use it." Other court rulings that could be used in test cases: a Maryland Court of Appeals decision in 1968 which ruled that the intended use of a car included "reasonably safe" transportation; and rulings by the California Supreme Court and the District of Columbia Court of Appeals that driving without the seat belt buckled has become a danger foreseeable by the manufacturers.

But the strongest case for the air bag—and, implicitly, the clearest indictment of our government guardians—can be constructed from the following study:

In 1979 the NHTSA decided to find out how the twelve thousand air-bag-equipped vehicles (mainly GM cars) then on the road were doing. The agency discovered that these cars had been involved in 205 serious crashes in which air bags had been activated. In these accidents 292 front-seat occupants had been affected. Of this number only two were severely injured and thirty-five sustained moderate injuries. Although the air bags functioned flawlessly in every case, five occupants were killed. One was a six-week-old unrestrained infant. Two died because they were in crashes so severe that they destroyed the entire front seat section of the vehicles. In the fourth case the cause of death may have been a heart attack. The fifth fatality occurred when the driver was ejected from his car after hitting a tree at one hundred miles per hour.

Two-thirds of these accidents were so serious that the automobiles were damaged beyond repair. Of the 292 front-seat occupants, 250 received minor injuries, or none at all.

Close-Up (2)
FORMALDEHYDE POISONING

"I thought I was going to lose my mind. The horrible headaches, coughing spasms, and my eyes . . . I was going crazy, not knowing what was going wrong with me or what was causing my problems."

—Helen Miller,
victim of formaldehyde poisoning

In 1977, with their heating bill climbing out of sight, Michael and Josephine Wagner did the sensible thing: They had $1,250 worth of urea-formaldehyde foam insulation—which in its liquid state has the consistency of shaving cream—pumped into the walls of their home in Toms River, New Jersey, where it quickly hardened.

Two weeks later, Michael Wagner collapsed and was taken to the hospital. He had been poisoned by formaldehyde and would henceforth suffer from sensitization to this substance, becoming dizzy around certain cleaning products and plastics. Cigarette smoke and permanent press clothing would also make him ill, and his wife would have to stop wearing cosmetics because of his sensitivity to them. Ultimately, Michael Wagner had to move out of his home and give up his job as a high-school science teacher. His daughter began to get headaches and breathing problems when exposed to felt-tip pens, and he had to ask her teachers not to wear perfume in the classroom.

His current situation, he explained, was "almost like living in a bubble, except the bubble can't be plastic, because that contains formaldehyde."

The Wagners were not alone; across the country several thousand other homeowners were suffering from the toxic effects of urea-formaldehyde insulation. Unknown to them, the regulatory wheels that would lead to the banning of this foam insulation were already turning, but turning very slowly.

Five years later, in February 1982, when the U.S. Consumer Products Safety Commission (CSPC) banned urea-formaldehyde foam insulation (UFFI) in homes because it posed an "unreasonable risk of injury from irritation, sensitization and cancer," the government and most citizens believed that their formaldehyde worries were over. They weren't—and they aren't.

In a shocking April 1983 decision, a federal appeals court in New Orleans overturned the CSPC's ban on foam insulation. And formaldehyde has many other uses in our society, which present what The New York Times has called an even "bigger headache" than foam insulation. The use of formaldehyde—a simple chemical consisting of nothing more than carbon, hydrogen, and oxygen—in wall insulation represents only a tiny part of its place in American life. In fact, it is found in products which (according to the Formaldehyde Institute, a trade group) make up 8 percent of the nation's Gross National Product, playing a part in such mundane items as toilet seats, antiperspirants, and rayon. It is used in manufacturing strong, lightweight plastics for automobiles. It also acts as a preservative in toothpaste, shampoo, and cosmetics, and helps make superabsorbent paper towels "super" and permanent press "permanent." In some of the products in which formaldehyde is used its presence is minimal. On the other hand, it is found in high quantities in cigarette smoke.

It is estimated that during 1983, the American consumption of formaldehyde probably exceeded 7.5 billion pounds, up from 6.4 billion pounds four years earlier. More than half of this went into the production of synthetic resins. Pressed-wood products are formed by mixing wood chips and sawdust with a bonding agent, and then compressing the mixture into sheets at high temperatures. Urea-formaldehyde resins are widely used as the bonding agents in such pressed-wood products as particle board, fiberboard, and plywood, which are probably found in a majority of American households. They also have a major place in the manufacture of mobile homes, which represented 25 percent of the new housing market (250,000 units sold annually) in the early 1980s. In conventional homes, pressed-wood products are used extensively in remodeling, in the form of kitchen cabinets, shelving, and paneling, and in furniture, including bookcases, dining sets, and dressers. Particle-board manufacturers sold 3.9 billion square feet of their material in 1981.

The problem with formaldehyde in pressed wood is the same as the problem with formaldehyde in foam insulation: outgassing. The urea-formaldehyde resin in the pressed wood breaks down and "gasses out" into the air for periods lasting up to five years. The amount that gasses out depends on many factors, and the effect the pungent, colorless gas has on people also fluctuates depending on the amount of formaldehyde present and the sensitivity of the individual. ("Medium density" fiberboard is by far the strongest emitter.)

"Outgassing," the Consumer Federation of· America contends, "represents a significant health risk to American consumers, and because of the ubiquitousness of [pressed-wood] products, poses a health threat of much greater magnitude than UFFI, which found its way into a smaller number of American homes."

But the degree of risk is open to debate. Some have called

for a total ban on all pressed-wood products containing formaldehyde, while makers of the chemical and others who find it integral to their products claim it causes no harm. The CPSC, still smarting from the overturn of its foam insulation ban, continues a drawn-out investigation of the problem.

The formaldehyde controversy began brewing in the mid-1970s. In October 1976, a few months before the Wagner family moved into their home, the district attorney's office in Denver, Colorado, filed a petition with the CPSC, requesting that the Commission develop a safety standard for certain types of home insulation products, including UFFI. It claimed that there was an unreasonable risk of injury from irritation and poisoning associated with the insulation.

Nearly two and a half years later, on March 5, 1979, the Commission decided to defer a decision on UFFI, but instructed its staff to evaluate further information on the topic. In press releases issued during the following year, the Commission warned that formaldehyde gas could cause health problems; that one study indicated it caused cancer in laboratory animals; and that (according to the National Academy of Sciences) exposure to the substance should be kept at the lowest practical level. But still the regulatory wheels turned slowly.

Formaldehyde poisoning, however, is only part of a larger problem: indoor pollution. A government study, released in March 1984, revealed that air pollution may be ten times worse inside American homes than it is outdoors. Among the hazardous pollutants found in forty test homes were asbestos, chloroform, and benzene. Formaldehyde was singled out as an increasing problem, with other major contributors including aerosol sprays, cleaning products, and appliances.

In June 1980 the CPSC published a proposed rule requiring potential purchasers of foam insulation to be notified of the potential adverse health effects associated with it. The CPSC solicited comments on this proposal, and sixty-six citizens and industry spokespersons responded. The Commission concluded nine months later that there was an unreasonable risk associated with UFFI, and that "no feasible standard applicable to the product could adequately reduce the risk." On February

5, 1981, the CPSC proposed a ban on the product. (Canada and two states—Connecticut and Massachusetts—had already banned it, having heard enough.) A petition from the Formaldehyde Institute asked the Commission to merely set safety standards for the product, but the CPSC in February 1982, by a four-to-one vote, decided to ban foam insulation formally.

In statements made after the vote, three of the CPSC's five commissioners blasted industry for implying that the commission had lacked integrity or expertise in the matter. R. David Pittle said that in his eight years with the Commission he had never seen such an "attack"—explaining that even though he agreed with the industrialists that a "zero-risk" society was an ideal "that neither corresponds to reality nor is consistent with the strictures of the Consumer Product Safety Act," he had voted for the ban because even at low levels, formaldehyde could have "significant adverse health effects."

Most dramatic in his official statement was CPSC Commissioner Sam Zagoria, who had previously opposed the ban. Zagoria called the ban "a climactic point for this tiny government agency," and said that finalizing it was a "far-reaching action taken only a few times in the Commission's [ten-year] history." He had finally supported it, he said, because, "The unhappy truth about this product is that you don't know it will hurt you until it is too late—until it is in your home. And then you can't just unplug it like some electrical appliance and remove it." Some installers, when asked for advice, had told suffering residents, Zagoria said, to "try ventilation, vapor barriers, dishes of ammonia, solutions of vegetable roots and leaves, or coffee grounds."

Zagoria clearly took to heart his responsibility, as one of only five people who held the fate of thousands in their hands. "What we do here today may decide whether twenty to twenty-five years from now a number of Americans will be struck down by and die prematurely of cancer," he said. "We know, too, that this cause of death is avoidable. There are alternatives to this kind of insulation. This one presents an unreasonable risk of injury. We need not add an extra load of formaldehyde to our daily living—there is enough formaldehyde from other products and processes."

Another commissioner, Stuart M. Statler, saw his responsibilities a little differently. While acknowledging that UFFI did indeed pose an "unreasonable" risk to consumers—and after reiterating an earlier vow that he "would not put the product in my home"—Statler said that a ban on the product would bankrupt too many businesses, devalue too many properties and deny consumers the right to choose what sort of insulation they wanted.

From the time of the first proposed restriction on UFFI in 1976 to the banning of the product in 1982, the foam had been installed in over 400,000 American homes, putting an estimated 1.5 million occupants in jeopardy.

The formaldehyde insulation industry did not take this startling setback in stride. "To just say formaldehyde is a hazard is ridiculous," said Josh Lanier, executive director of the National Insulation Certification Institute, a trade group of UFFI manufacturers. "We've been using this chemical for 100 years. There's no magic whereby in 1982 it suddenly becomes dangerous." Lanier also pointed out that among 400,000 homes with UFFI, the CPSC dug up only 2,000 complaints. "Compare that to something like chain saws, where the CPSC found 86 deaths and 123,000 injuries in 1981," he said, "and you'll see how little sense the UFFI ban makes."

Health experts pointed out that the analogy to chain saws was not quite appropriate, because fatalities related to their use were usually immediate, whereas deaths from UFFI might increase within ten to twenty years if its potential for causing cancer were realized. But Lanier had good reason to be angry: his industry had just been handed its death notice. At the time of the ban, half a dozen major firms were manufacturing UFFI, and 200 to 300 companies, with an average of eight to ten employees each, were installing it around the country. (Adverse publicity surrounding the product had already caused its share of the home insulation market to slip from 5 to 1 percent.)

Even before the ban, lawsuits had been a problem for the industry. Hundreds had been filed, including a $2 billion class-action suit on behalf of 70,000 to 130,000 residents of New York

State. Several six-figure settlements had already occurred, the Wagner family, for example, having won $225,000 from the manufacturer and installer of its insulation. As more suits approached trial, the big question was whether the manufacturers knew enough about the health hazards their products posed to have warned consumers about them. The settlements would rise substantially, in the form of punitive damages, if the manufacturers were found negligent.

For some who had already been driven out of their homes by the fumes, the choice was clear: nothing could make them move back; they had left forever, hoping to make up their losses (and then some) in court. Others had had the foam ripped out of their walls (at a cost that ranged from $6,000 to $20,000). But most simply stayed put, shunning even the smallest alteration.

Those who left the UFFI undisturbed saved on remedial costs but had to pay another price: the devaluation of their homes. Some real estate agents refused to even list homes insulated with UFFI; others made sellers state in writing that potential buyers should be aware of the hidden presence of foam insulation. Many dream houses had turned into white elephants.

"It is really too soon to tell how the ban will affect home values," said Bernard Goodman, a former president of the American Society of Appraisers. "You are dealing with the psychology of buyers. Maybe it will be like saccharin, where people eventually say it wasn't as bad as we thought."

Of course, relative affluence had a lot to do with who stayed and who fled, who renovated and who stood pat. And the level of risk in different homes varied. Experts agreed that the first year of installation was the prime period of danger, with the fumes decreasing rapidly (but not always totally) thereafter. Some walls were more porous than others; a lack of ventilation in some homes made them formaldehyde-traps; heat and humidity greatly increased outgassing. Following the UFFI ban, thousands of homeowners with foam in their walls paid from $50 to $500 to have local laboratories test the formaldehyde levels in their homes, or took their own readings with badges ("passive monitors") made by the 3-M Company. In most cases very low levels of formaldehyde were found. (After testing the performance of "passive monitors" used in some ten thousand American homes, the National Institute for Occupational Safety and Health [NIOSH] later declared that they were unreliable.)

Then in April 1983 the U.S. Court of Appeals for the Fifth Circuit, in New Orleans, set aside the urea-formaldehyde ban. The court acknowledged that the product was "not completely innocent" and that consumer complaints "do identify a real problem," but it found that the CPSC had based its decision on inadequate evidence and improper regulatory procedures that had denied the foam insulation industry a fair hearing. The court specifically criticized the validity of the commission's case against UFFI as a cancer risk.

"Out of the clear blue sky," says Harleigh Ewell, a CPSC attorney, "they said a study of 240 rats was not enough, when most experts say 100 is plenty." Ewell criticized the notion of courts making an "independent assessment" of technical data, but says that the basic reason for the overturn was that the court felt there wasn't enough evidence of risk to warrant the cost of a ban. "We were admittedly unable to quantify the symptoms to justify" a risks-versus-benefits analysis, Ewell explains.

The CPSC asked Rex E. Lee, the Solicitor General, to appeal the circuit court's decision to the Supreme Court, but on August 25, 1983, in another move, the Justice Department announced it would not carry out the appeal. In a written statement, Lee gave no reason for this decision, but his repudiation of the CPSC drew an unusually strong comment from its chairman, Nancy Steorts—another Reagan appointee. She warned consumers that they should not be led "to the mistaken belief that this product is now considered safe." And David Greenberg of the Consumer Federation of America, citing "monumental" evidence that the insulation was indeed unsafe, said: "The ban on insulation was the one bright spot in the health and safety area in the Reagan years, and they have wiped it out."

Would the overturn of the ban allow the UFFI industry to revive? "The unhappy news is that foam insulation may be creeping back into the market," Sam Zagoria noted. "The insulation has not changed; it still threatens users. . . ." Senator Carl Levin of Michigan was so concerned that he introduced a bill that would end the federal tax credit for homeowners insulating their homes with UFFI. ("Brutal as it sounds," responded an I.R.S. spokesman, Steven Pyrek, "we are not concerned over health hazards.") The CPSC issued a formal warning that the product was still "hazardous." The formaldehyde industry, nevertheless, viewed the overturn of the UFFI ban as crucial in aiding its efforts to clear the name of formaldehyde so that its use in hundreds of other products would not be threatened.

Maureen Tiderman, a twenty-four-year-old school teacher in Port Angeles, Washington, bought a $19,000 Fleetwood mobile home in 1977, while training for a mountain-climbing trip. After living in the home for six weeks, she developed chronic asthma and a sensitivity to air pollutants, and had to keep emergency oxygen supplies in her home, car, and classroom. "I can't go climbing anymore," she complained, "and I can't stray far from my oxygen." In 1981, as the result of a damage suit, a jury awarded Tiderman $566,000.

Meanwhile, in Moultrie, Georgia, Helen Miller was using an air filtration system (commonly referred to as a "bubble") to aid her breathing—a need that she believed came from years of exposure to fumes emanating from particle-board flooring she had installed in her home in 1975. For four years "my life was hell," Miller recalled. Her illness was not diagnosed until 1979, after she saw a television program on formaldehyde poisoning.

Maureen Tiderman and Helen Miller now have a lot of company in court and in doctors' offices. Several thousand people have complained to the CPSC about health problems—ranging from short-term discomfort to long-term impairment—that they believe are linked with formaldehyde fumes leaking from pressed wood and particle board in their homes. Hundreds of lawsuits have been launched, with plaintiffs winning judgments in a high percentage of them. The United Auto Workers Union has sued the U.S. Labor Department for concealing evidence that formaldehyde endangers the health of workers. Other union health and safety activists have urged a ban on formaldehyde in beauty and barber shops.

Perhaps more than any other single person, Connie Smrecek is responsible for this citizen offensive against formaldehyde. In February 1977, the Smrecek family remodeled their Waconia, Minnesota, home, using particle board. Within a year, Connie later reported, her entire family had developed respiratory problems and stomach disorders. One of her sons suddenly began experiencing learning disabilities, and she herself became sensitized to several consumer products.

So, in 1979, she started an organization called SUFFER (Save Us From Formaldehyde Environmental Repercussions), which now has chapters in thirty states. Since then, Connie Smrecek has been contacted by more than ten thousand strangers. Some wonder whether they should rip foam insulation out of their walls; others suspect that their health problems are tied to the paneling in their mobile homes; others, like Connie herself, believe that recent remodeling in their homes is haunting them. Like many homeowners, they once believed that their home was their castle but, as Connie says, "something in their castle made them sick."

Connie claims she has gotten up to two hundred calls and letters a day, which explains why she now uses a computer to store information. She's been contacted by morticians, who use formaldehyde in embalming fluids; and beauticians, who use it to disinfect combs and other objects. She's heard from clerks in department stores who feel they are affected by the formaldehyde in clothing; from workers in lumber yards; and from hospital workers, who use formalin, a type of formaldehyde, in dialysis machines. (An estimated two hundred fifty thousand American workers in all are exposed to formaldehyde.)

"A lot of people," she says, "have trouble convincing a doctor or an attorney of their problem because it sounds incredible." She points out that "Formaldehyde is so dangerous because it is a sensitizer, which is not true of most chemicals. Most people who have allergies also have problems with products containing formaldehyde—in many cases the allergies may have begun with sensitization to formaldehyde."

Moreover, she notes that "A lot of people have become sensitized to formaldehyde but . . . don't know it, so when they use a certain shampoo, or eat a particular food, or someone comes to their home with tobacco smoke on their clothing, and they get sick, they don't know what's wrong."

SUFFER has gone so far as to suggest that formaldehyde may be a major cause of the sudden infant death syndrome (SIDS). This view is not shared by many in official positions. For example, Joan Jordan, an inspector with the Connecticut Department of Consumer Protection, feels that the formaldehyde threat in homes has been "blown out of proportion." When people call for advice, Jordan tells them that if they've had formaldehyde-laced products in their homes and haven't had any health problems, they won't have any in the future. (Callers who report illnesses are another matter, says Jordan.) But one of SUFFER's attorneys in Minneapolis, Allan Shapiro, feels that the formaldehyde problem, rather than being a case of mass-hysteria, simply stems from greater public awareness that it exists.

Formaldehyde has not been firmly regulated in the marketplace. In 1982 the Environmental Protection Agency declared the substance a "low-level risk," and decided that the Agency would not regulate the chemical. Representative Albert Gore of Tennessee termed the decision against regulating formaldehyde "crass, calculated, cynical."

In May 1984 the EPA announced a change in policy concerning the regulation of formaldehyde. The agency said it would give a priority to considering the regulation of the substance as a possible cause of human cancers. The options under consideration by the agency range from doing nothing to a ban or partial ban on uses that might place a significant number of people at risk.

The EPA's decision to reconsider its 1982 policy position was influenced in large measure by suits brought against it by the Natural Resources Defense Council and other environmental groups. The agency also said that its change of position was influenced by the results of laboratory studies showing that formaldehyde causes cancer in rats at high levels of exposure.

The Formaldehyde Institute publicly characterized the EPA's May 1984 decision as inappropriate. "When all the available data are carefully assessed, the agency and the public will agree that formaldehyde's uses do not present any significant risk to health," said John F. Murray, president of the Institute.

How real is the danger? The Formaldehyde Institute says flatly that "off-gassing associated with formaldehyde-resin products used by consumers does not produce levels of formaldehyde that are considered harmful," and that "any potential irritation effects of off-gassing can be reduced in the home by opening windows or providing for increased ventilation."

But is there a "safe level" for this substance, below which it does not bother, harm, or sensitize people?

Currently, no national standards exist for allowable levels of formaldehyde in what is often referred to as "residential air." The U.S. Department of Housing and Urban Development (HUD) in 1983 set a standard of 0.2 ppm for plywood and 0.3 ppm for particle board, stating that "standards resulting in lower emission levels are not cost-effective."

The Consumer Federation of America (CFA) has petitioned the CPSC to set a standard of 0.1 ppm of formaldehyde for all homes, while the Formaldehyde Institute quotes studies showing no negative response to the substance (besides eye and nose irritation) below 3 ppm.

When referring to exceedingly small quantities of a substance in the air—so small that they can be measured only in millionths—it may seem that small differences in these quantities would be irrelevant. Yet the health of millions of Americans, and the future of the pressed-wood industry in America, may ride on those differences and how they are resolved in Washington.

Arguing for the 0.1 guideline, the CFA has pointed out that it is the national indoor standard for formaldehyde in the Netherlands, Austria, West Germany, and Norway. The National Academy of Sciences has noted that about 20 percent of healthy young adults may respond unfavorably to formaldehyde at levels of less than 0.25 ppm, and that the percentage of persons so affected is probably higher in the general population, which includes groups more susceptible to the effects of the chemical. (The NAS estimated, for example, that from 10 to 12 percent of Americans may have "hyper-reactive airways" that could make them more susceptible to formaldehyde's irritant effects.) The Academy, in fact, failed to find any level that could not cause adverse effects. On this basis, its Committee on Toxicology recommended that formaldehyde in indoor air be kept at the lowest practical levels.

But what does "practical" mean? Do you gear standards to the "healthy" or the "susceptible" individual? And what do you do about the almost infinite sources of formaldehyde?

It would be hard to find a home in America where permanent-press products, carpets, cosmetics, or cigarette smoke—to name just four formaldehyde-rich products—are not contributing to indoor air pollution. "It is impossible to prevent all exposures to formaldehyde," says Dr. Richard A. Griesemer, director of the Biology Division of the Oak Ridge National Laboratory (and chairman of the Federal Panel on Formaldehyde). "To prevent the toxic and carcinogenic effects of formaldehyde, however, it is desirable to avoid adding excessive exposure to those that cannot be avoided." Griesemer recommends that "human exposure not exceed 0.1 ppm for any extended period of time."

The problem with this is that in many homes the "background level" of formaldehyde in the air—not counting the amount contributed by outgassing from pressed wood—already approaches the 0.1 ppm standard. Moreover, this background level has probably risen in recent years due to attempts by many homeowners to "seal off" their homes to save fuel costs.

Because this background level is so significant, the Consumer Federation petitioned the CPSC to limit outgassing from pressed wood to 0.05 ppm of formaldehyde. However, the pressed-wood industry, noted the CFA in an August 1982 petition to the CPSC, "has fought every attempt to establish safe levels."

While the formaldehyde controversy continues and the CPSC contemplates further action on both foam insulation and pressed wood, many mobile-home manufacturers are turning to gypsum as a substitute for plywood, while school laboratories are ordering animal specimens embalmed in "alternative" preservatives. Makers of pressed-wood products are attempting to reduce outgassing by applying substances (such as ammonia) to particle board. They have also developed what they call "low-emission" plywood and particle board, which cost about 5 to 20 percent more than the ordinary versions of these materials and reportedly yields a smaller amount of outgassing. More-over, although the pressed-wood industry, as the CPSC has noted, "would like to continue using U.F. resins because of [their] low costs," manufacturers are examining several alter-native binding agents.

Despite this, notes the CFA, "none of these methods alone can currently reduce emissions to the desired .05 ppm" level. Yet somehow, says the CFA, industry must find a way to reach this standard, because "the health of the American public is at stake." If the standard cannot be met, the Federation has said that it will support a federal ban on the use of urea-formalde-hyde resins in pressed-wood products.

In responding to this threat, two attorneys for the formal-dehyde industry, Edward Canfield and John R. Gerstein, have warned that "whether the technology was developed at great expense or whether a ban was imposed instead, the result would have severe economic impacts not only on industry, but on the public as well. It would, for example, drastically increase the price of manufactured housing, thereby depriving many con-sumers of the last vestige of affordable, low-cost housing. When one balances the scant and unreliable health data against the enormous financial burden on the public and industry, it is clear that a product standard such as that suggested by CFA is manifestly improper."

From the time when the CPSC received its first UFFI-related complaint to the day a ban on the insulation was instituted, over six years passed. Because of the widespread use of pressed wood and the relatively low levels of formaldehyde it emits, any significant action on the outgassing problem will probably take even longer.

Is a quicker response called for? As long ago as December 28, 1982, Ronald L. Medford, project manager of the CPSC's Household Structural Products Program, in an internal mem-orandum, referred to the reports of adverse health effects from consumers exposed to pressed-wood products as "a substantial number." The number at that point was 2,100 complaints—five hundred more than were on hand when the CPSC proposed its UFFI ban. And it has climbed since.

"Industry pleads again for more time for study, and this is

an alluring plea," said the CPSC's Sam Zagoria, in explaining why he had finally voted to ban UFFI in 1982. "There would be few victims visible to object. Cooperation with industry in today's society is essential, but cooperation is not to be confused with abdication of responsibility. . . . As has been said, 'Inaction has considerable costs of its own. . . .' While we study charts of costs, benefits, risk assessments and the like, we are talking about real people and how to help them avoid premature death and unnecessary suffering. . . ."

6
Calculating Risks

The practice of putting into numbers the value of such intangibles as clean air and good health reached some kind of nadir in the mid-1970s, when the president of a Cambridge, Massachusetts, think-tank attempted to come up with a "net psychological quality-of-life cost of mastectomy." Looking for a single figure that would include all the costs to the patient of a radical mastectomy, including its effects on a woman's "sex life," he decided that the value of a single "sexual experience" was roughly $40, "and if we assume an average frequency of fifty per year, the annual shadow price is $2,000." But he conceded that this "cost" could be discounted by "the probability of a satisfactory sexual adjustment.

Offensive as this calculation may be, it is only a natural manifestation of the current trend to put a dollar value on everything, feed the figures into a computer and wait for an "objective" answer to a problem to pop out. The machine does the thinking, untainted by ideology, politics, or compassion. This may seem cold-blooded and a shirking of our responsibility as analytical animals, but (as we have seen) it is perhaps no more flawed than all the other extremely subjective ways in which we choose the level of risk at which to take protective action.

The problem with cost-benefit analysis is that unlike the other forms of analysis, it *appears* neutral; down deep it is often as biased as the opinion of the most subjective expert or bureaucrat.

Usually considered a new phenomenon—some commentators seem to believe that Ronald Reagan invented it when he underlined its role in federal regulatory decision-making in 1981—cost-benefit analysis has actually been around for many years, dating back at least to the 1930s, when the U.S. Army Corps of Engineers drew up equations to substantiate its requests for more money to build dams (some people claimed the Corps routinely overestimated the benefits of the dams). It has been used by economists for years, written into certain laws, and utilized in official and unofficial ways by several presidential administrations preceding the Reagan Administration.

Actually, this kind of comparative thinking is almost second nature to us. Every time we choose whether or not to drive our car faster than the legal speed limit, we are doing a risk-benefit analysis, weighing how much faster we will get where we're going against the odds of our getting caught, and whether our speeding will increase the chance of an accident.

In the past few years, however, formal cost-benefit analysis has come into its own. Some of its practitioners trace its birth to 1969, when *Science* magazine published Chauncey Starr's classic paper "Social benefit versus technological risk." Risk analysts now have their own societies, academic niches, and institutes. In fact, the notion of making decisions on the basis of pluses versus minuses has become so acceptable that politicians apply it to the most far-flung endeavors in an effort to explain seemingly implausible actions. (After the United States dispatched troops to Lebanon in September 1982, for example, Vice President George Bush admitted to television reporters in New York City that it was a risky step, but added that "there's a risk any time you cross the street here. We weighed the risks very carefully," explained Bush, "and decided that the benefits outweighed the risks.")

No matter what its critics say, there is a place, even a need, for cost-benefit analysis in modern society. Health hazards are proliferating—or at least we are more acutely aware of them. Yet resources to eradicate or prevent these hazards are painfully finite and perhaps shrinking. Observing that "risk-benefit [analysis] assumes a world in which many justifiable actions cannot be taken," science writer Fred Hapgood explains that "we have to choose which hazards to go after, and at what resource level. Hard choices must be made."

The bulk of funding for many of these efforts comes out of one large pool—the federal budget. Those who set budget outlays must first decide which threats to deflect, and with how much money. This already constitutes two decisions. Once a threat has been identified, the choices for dealing with it abound. Do you try to reduce heart disease, for example, by pouring money into pollution control, basic research, antismoking commercials, or mobile cardiac-treatment units? Computing the costs of running each program, and estimating the results of each, can indicate where the "most bang for the buck" can be had. "The notion," as Murray Weidenbaum, a top economic adviser to the Reagan Administration, put it, "is that if you examine the consequences of your actions you get better decision making."

Cost-benefit analysis does work well when it involves deciding where limited funds will be spent to attack a single problem—to save lives in one universe. Difficulties arise, however, when the analysis leads to a decision that causes funding to *leave* that universe, such as when someone decides that saving lives at $500,000 per person isn't worth it, and that the money could be better spent on a new battleship.

Cost-benefit analysis got a tremendous boost when, three weeks before he left office, President Gerald Ford, in an effort to stem inflation, issued Executive Order 11821, requiring federal agencies to conduct cost-benefit studies every time they proposed regulations that would have strong economic impact. President Carter renewed the order, and one month after taking office President Reagan followed suit, making his Office of Management and Budget the arbiter of regulations. Critics charged that this transformed cost-benefit analysis from a useful aid to an absolute requirement.

Reagan felt that his landslide election victory the previous November was a mandate from the public to cut government spending and get Washington off the backs of businessmen everywhere. His executive order said that "regulatory action shall not be undertaken unless the potential benefits to society from the regulation outweigh the potential costs to society." Before issuing any new rule, federal agencies were to weigh its potential costs and benefits, including those that "cannot be quantified in monetary terms." The agencies were then to choose the regulatory approach that involved "the least net cost to society."

This set off a spirited debate that touched on politics, economics, and philosophy. Proponents of the Reagan rule argued that it would redress the long-standing practice of government spending of billions of dollars on measures that returned (in terms of improved health, safety, and other benefits) only a fraction of the outlay. Opponents of the rule felt that the Administration would simply use the cost-benefit tool as a club to beat to death regulations it didn't like. They questioned the utility of this tool and the intellectual integrity of those who relied on it.

Thus, while Murray Weidenbaum called cost-benefit analysis a "neutral concept," Ida R. Hoos, a research sociologist at the University of California at Berkeley, said that it was "about as neutral as asking a fox into the henhouse to observe the color of the eggs. There is nothing magic or scientific about it," explained Hoos. "It is almost always an *ex post facto* justification of a position already taken." However, Christopher C. DeMuth, who supervised regulatory affairs at the White House's Office of Management and the Budget (OMB), asked: "What's the alternative? Flipping a coin? Consulting a Ouija board?"

The use of the cost-benefit weapon by both sides in the debate over renewing the Clean Air Act (see following "Close-Up") would receive a good deal of attention, but its everyday use by federal regulators during the same period went pretty much unexamined. When the Labor Department, for example, published a new set of regulations on labeling toxic chemicals in the workplace, it said that the rules would reduce cancer deaths by four thousand annually. The OMB challenged this, saying that the regulations would prevent "only four hundred" cancer deaths a year—too few to justify their costs to employers and chemical manufacturers. The workers' right-to-know about toxicity, the OMB said, "should not be considered a 'right' in isolation from the cost considerations" this would entail.

An even larger segment of the public was affected by the Reagan Administration's decision to reduce the ability of car bumpers to withstand bumps.

A 1979 regulation required that its front and rear bumpers protect a car from any damage in crashes at up to 5 miles per hour. This increased the cost of a new car slightly, but resulted in fewer trips to the repair shop and lower insurance premiums.

Now the NHTSA has decided to roll back this bumper standard to 2.5 mph—or "about the speed of a person walking," according to the Consumers Union. Because federal law requires that the bumper standard achieve "the maximum feasible reduction of costs to the public and to the consumer," the NHTSA had to show that the benefits of a rollback outweighed the costs.

The Consumers Union reported that in all its years it had "rarely encountered a change as dramatic" as this one, and observers agreed that this was a classic test for Reagan's new cost-benefit requirement. There was no need in this case to estimate such sticky values as clean air and human life; this was dollars-and-cents accountancy at its clearest. NHTSA analyst Barry Felrice said that the bumper question seemed "almost perfectly suited" to a cost-benefit study.

After completing its study, the NHTSA, in July 1982, did roll back the standard. But supporters of the crash standard were unconvinced by the agency's arithmetic, and took the agency to court.

The NHTSA claimed that the bumper standard was costing consumers nearly $300 million a year: $28 per car times 10 million cars sold annually. It broke down this way: using a lighter bumper than the then-current model would produce savings of about $93 per car in production and fuel costs (over the ten-year life of the car). From this total the agency subtracted an estimate of what the stronger bumpers saved in repair costs ($65), and came up with the $28 net gain per car for the new bumpers.

The problem with this, insurance spokesmen pointed out, was that insurance rate increases for cars with "creampuff" bumpers (as Ralph Nader described them) would more than wipe out the production and fuel savings. Others observed that mixed in with NHTSA's factual figurings were some pure guesses. The agency, for instance, placed a $14 value on the "aggravation" for motorists of getting repairs—including the cost of making calls, stopping by the service station, and going without a car while the repairs were made. This seemed empirically wrong, and a survey for the Center for Auto Safety revealed that motorists themselves put the "inconvenience value" at somewhere between $50 and $200. A small matter in the real world, perhaps, but mammoth in cost-benefit land, where one small adjustment on either side of the ratio can doom or justify proposed regulation.

Seven months after the new standard went into effect, hearings were held on Capitol Hill. The Insurance Institute for Highway Safety, which had conducted crash tests on cars with new and old bumpers, found that the repair costs with the new bumpers were sharply higher than accounted for in the government's equations. In one 5 mph crash involving a Volvo, the repair costs jumped to $745 with the 2.5 bumper from $143 with the old bumper. A crash involving a Honda Accord showed costs increasing from zero to $299. At the same time, there was no evidence that the sticker price of new cars had been reduced by the $30 to $35 promised by auto makers because of the cheaper bumpers. Senator John Danforth of Missouri, a Republican, called the new bumper standard "fraudulent" and added: "The Government should have an adequate standard or have no standard at all."

Yet the 2.5 mph standard held, to the amazement of its critics. "The only explanation I can come up with," commented Clarence Ditlow of the Center for Auto Safety, "is planned obsolescence. If you have poor bumpers on a car, clearly they're going to get banged up more, and people will buy new ones." The Center disclosed that about 60 percent of 1984 model cars used the weaker bumpers, with Ford the leading proponent of the tougher versions. None of the manufacturers have publicized which bumpers they use and the Center said that in most cases car buyers can't tell the difference in the showroom.

"Neutral" as it seemed on the surface, the bumper question turned on several issues of cost-benefit interpretation. An unnamed NHTSA official confided to the *National Journal*: "You could pretty much throw a dart and show any of the alternatives to be the best," and added that this "shows how difficult it is to base a decision on a rigid cost-benefit analysis." Yet compared to many issues involving more serious health and safety questions, the bumper standard was a piece of cake.

At the heart of the controversy over cost-benefit analysis are two questions: How do you put monetary values on the invaluable? And who decides which risks and benefits are considered in these calculations?

"The cost of a thing," Henry David Thoreau wrote in *Walden*, "is the amount of what I will call life which is required to be exchanged for it, immediately or in the long run." By this yardstick, most people, if asked to put a price on their own life, would either call it infinite or contend that it is impossible (or even immoral) to attempt to do so. Yet the average American has been letting government officials do this for decades, at first informally, and now with increasing audacity. In some cases experts pluck a "life-worth" figure out of the air (for some reason the figure $250,000 reappears frequently), although the NHTSA arrived at the more precise figure of $287,175. (This was several years ago, with recent inflation not factored in.)

Even the most serious estimate can never, of course, measure the personal cost of a lost life as viewed by the unfortunate individual who is about to expire. Analysts do factor in a dollar value for some things that can be measured, such as hospital, funeral, and insurance costs. What cost-benefit analysts seem most interested in, however, is how much the death of one productive member of the work force will cost the economy. Attempts to assign an economic weight to life are often based on a "human capital" approach, which puts the value of saving a life in terms of the economic contribution of that life to society. It assumes that programs that reduce death and suffering have as their objective the maximization of the gross national product. From this purely economic perspective, women are less valued than men, retired people and artists less valued than workers, and low-paid workers less valued than well-paid workers. The human-capital approach can also lead to the conclusion that death is preferable (GNP-wise) to a disability that may involve expensive medical treatment on top of an earnings loss.

A study done for the New York State Department of Health in 1979 concluded that estimates on the value of life "range from $49,226 to $1 million with most values between $200,000 and $300,000." The researchers said that these estimates, along with such factors as a person's income, productivity, and the cost of treating the disease that might affect that person should be taken into account in determining whether the state should spend money to reduce the number of cancers caused by chemicals. "Despite justifiable moral and philosophical objections," they said, "this dilemma must be confronted if economic assessments are to be made." However, the researchers acknowledged that this methodology would "undervalue lives of housewives [and the] elderly, unemployed, and underemployed," and that it also did not incorporate "social values or the utility of life to an individual." The report dryly noted that several extra deaths across the state due to chemical exposure would probably not cause public concern but "these events certainly would be noticed by [those persons] directly affected, their friends, relatives, and business associates."

Most often, however, analysts peg the price on life by examining the amount of money spent in certain programs to prevent deaths. If this does not reveal the true value of life, it at least reveals regulators' *opinions* of its value. For example, a table put together in 1981 by William Nordhaus, an economic adviser to President Jimmy Carter, showed that society seemed willing, to save a life, to pay $166,000 for a kidney transplant and $621,000 for long-term dialysis treatments in the hospital. A lawnmower safety standard proposed by the Consumer Product Safety Commission would have saved lives at a cost of $390,000 to $3,120,000 each. Nordhaus's analysis of such rules led him to identify a social value on each life saved of $480,000.

Yet another way to look at the value of a life is to consider what judges and juries have awarded plaintiffs in damage and malpractice suits. It is hard to justify a $250,000-per-life ceiling when one considers (to quote some random, recent awards) a $1.75 million award to a cancer victim whose mother took the drug diethylstilbestrol (DES); a $7.2 million award in a case involving aggravated rape; and $11.2 million to the parents of an infant who suffered brain damage in his crib. And none of these victims died.

It might be argued that these huge awards were products of overemotionalism—but we tend to judge our lives and those of people we care about in a highly emotional way, and if $7 million is too much to spend to save a life, how, exactly, do we determine what to spend?

There is also the what's-it-worth-in-the-marketplace approach. Suspecting that the old saw that the human body was worth only from 98¢ to $5 was outdated, Dr. Daniel A. Sadoff, a Seattle veterinarian, did a quick computation in 1983 and came up with a new figure: $169,834. This was based on what wholesale chemical distributors would pay for only seven of the dozens of chemical substances in the body. (It didn't even count the $1,200 worth of blood we each possess.) According to Dr. Sadoff, for example, there are 140 grams of cholesterol in the average 150-pound person, with a wholesale value of $525.

If there is no agreement on the price of life, there is scarcely more of a consensus on the value of remaining free from a crippling injury or grievous illness. How do you judge the dollar cost of not getting emphysema? If a life is worth a quarter of a million dollars, is remaining free of brown lung disease worth $125,000? $50,000? A settlement plan drawn up by the Manville Corporation that would offer "fair, efficient compensation for all asbestos claimants"—the company is being sued by thousands of its former employees—suggested $45,000 as a fair payment to the average lung cancer victim.

Those who devise cost-benefit models rarely try to determine the value of these immeasurables. Instead they concentrate exclusively on things they can measure. Health defects, when they are not misjudged, are simply overlooked. At this point, it is hard to say which is more appalling: trivializing death and disease or ignoring it.

It is also difficult to measure such "benefits" as convenience, pleasure, and peace of mind. What is the monetary value of being able to drink clean water from your tap rather than having to drive to a nearby spring? Of enjoying a single, smogless summer day? Of being able to use an artificial sweetener in your coffee instead of sugar?

These are, literally, the $100 billion questions. When the EPA did a cost-benefit equation in early 1984 on a proposed revision in the Clean Air Act, it suggested that the net benefits (after deducting the costs) ranged from a loss of $1.4 billion to a gain of as much as $110 billion—depending on the value assigned to such things as human health and a cleaner environment.

Matters are no more precise on the "cost" side of the ratio. Costs can include everything from maintenance to legal fees. Critics accuse industry analysts of puffing up these costs to tilt any equation in their favor (i.e., *against* regulatory action). Mark Green, former head of Ralph Nader's Congress Watch, says that cost-benefit analysis is "institutionally biased against health-safety regulation" because cost data are controlled by regulated industries.

A strong manifestation of such bias appeared in 1974. Chemical industry studies asserted that all plants making polyvinyl chloride would close, with a cost to the economy of about $90 billion, if the OSHA adopted a strict standard on workplace exposure to polyvinyl chloride—a chemical linked to liver cancer. According to former Secretary of Labor Ray Marshall, the real cost of the standard, once it was adopted, turned out to be between $127 million and $182 million, and, said Marshall, "the industry has managed not only to comply with the standard, but to devise more efficient and profitable technologies that actually increase production at the same time."

Of course, both sides in these struggles are prone to come up with numbers that support their own arguments. Thus, estimates of the costs to industry of a regulation usually come from the industry that is to be regulated; estimates of the benefits are likely to come from the people who would benefit from the regulation or officials who favor it.

Such estimates are further compromised by the fact that both sides often base their numbers on studies carefully selected to support their views—studies that often conflict, since the choice of a scientist, expert, and statistics on which to base an argument is rife with bias. The computers which do the final cost-benefit accounting may not "lie" but they can only judge facts fed into them; they're only as good as their material. (The expression for this in computerese is: "Garbage in, garbage out.") However neutral computers may be, they can never produce an "objective" answer from highly subjective data.

Further complicating matters is the fact that in weighing risks, the threat to *humans* is often derived from laboratory experiments involving *animals*. Effects on animals must be projected onto man: a useful but horribly imprecise process.

And then there are the political facts-of-life that often enter into the equation. After the Reagan Administration in early 1984 put off a program to control acid rain, for instance, Senator Daniel Patrick Moynihan of New York suggested that EPA chief William Ruckelshaus had weighed the electoral votes in the polluting states against the electoral votes in the states affected by acid rain. Ruckelshaus admitted that such political considerations had been discussed but had not been the *basis* of the decision.

These biases, imprecisions, assumptions, and manipulations help turn serious cost-benefit analysis into just another numbers game whose impressive-looking formulas should be taken, in many cases, with several grains of salt. Opposing sides, using the same or quite different data, can easily present to the public a plausible argument for their own case. Instead of one cost-benefit "ratio," there are at least two; instead of one well-balanced answer we got a teetertotter.

Cost-benefit analysis is beset by a variety of other problems. Some, for example, feel that it is inherently unfair because the benefits of a product or substance are frequently enjoyed by only one segment of the population while the risks are inflicted on others. In other words: those getting the benefits often don't run the risks while those who run the risks don't get any benefits. This is especially true in the chemical industry, in which workers and those living in the vicinity of a plant often bear the risks while those living far away reap the benefits of the product.

Others fear the nightmare potential of the cost-benefit practice. Medical costs have skyrocketed to the point where already some are speculating that one day we may set rules that will deny certain kinds of medical care to the elderly so that resources can be better applied to the young. (This would let the gross national product play God.) Cost-benefit analysis is blind to human rights, according to Steve Kelman of Harvard University's Kennedy School of Government. He compares some of its uses to the doctors in the novel *Coma* who killed healthy people to use their organs to save other people—and defended themselves by saying, "We saved two people for every one we killed."

In a *New York Times* article in May 1983 entitled "Longer Lives Seen as Threat to Nation's Budget," Dr. Leonard Hayflick, director of gerontology studies at the University of Florida, revealed that few biomedical researchers are working on increasing the life span. Such work, he said, may not even be desirable because an increase of ten to twenty years in life span could have "absolutely catastrophic" social effects. At the same time two economists with the Federal Office of Management and Budget, Barbara Boyle Torrey and Douglas Norwood, warned that improvements in mortality rates would increase the "already ominous" growth potential in programs for the aged. One study calculated that if all the people who died prematurely in 1978 from heart disease had been able to reach their full life expectancy, the net loss for the federal government would have been $15 billion.

If resources *are* truly finite a finely focused cost-benefit approach *can* work well for the common good. The Federal Highway Administration, for example, examines the cost-per-life-saved of removing defects in highway design, ranks them in order of the greatest number of lives saved per dollar, and then encourages state highway departments with limited budgets to make their road improvements according to that ranking. (Thus, making guardrail improvements costs only about $34,000 per life saved, while creating "impact-absorbing" roadside devices comes to $108,000 per life.) This means "buying lives" in the most cost-effective manner, and few would object to that. Comparative analysis, notes William W. Lowrance, can help set priorities that are "defensible," avoid frittering away "worry-capital" on very small hazards, and concentrate attention on middle-range hazards that affect many people in significant ways and are subject to change.

Seen in another light, however, it means that some lives are too expensive to save. If we agree that society should limit its spending in preventing needless highway deaths, we accept the unavoidability of thousands of people dying on the highways. But what if we determine that money should be directed away from another sector of the vast budget to ransom these lives? Is it more important, for example, to launch another space shuttle or pump money into mobile cardiac units? Cost-benefit analysis can lull us into accepting what is instead of challenging us to consider rearranging the big picture.

Little wonder, then, that those engaged in constructing cost-benefit equations are self-conscious about their work. Steven Rhoads, a professor of government at the University of Virginia, has even argued that the practice should keep as low a profile as possible. "As for candor," he says, "I would tolerate a little dissembling in this area. . . . Admittedly, the absence of total candor leaves the way open for the periodic appearance of ambitious reporters and politicians who expose the Dr. Strangelove-like analysts at the heart of the bureaucracy." But, continues Rhoads, "this cost seems tolerable in an area where simple enlightenment is not to be expected or wished for."

By contrast, it would seem that the process of cost-benefit analysis should become more open to the public, so as to insure that individual concerns are being evaluated in an equitable manner. William Ruckelshaus seemed to recognize this when he became head of the Environmental Protection Agency in 1983. He suggested creating a national commission to find "some universal way" to measure health risks and assess how much society is willing to pay to reduce those risks. Noting that public opinion had become polarized on this issue, Ruckelshaus said that the commission might include poets, historians, and "people from every walk of American life."

This was, at last, a high-level admission that judgments about the amount of public pain and suffering that can tolerably be exchanged for a reasonable standard of living are too important to be left to the generals of cost-benefit analysis.

Close-Up
THE CLEAN AIR DEBATE

"I was elected to go to Washington to rewrite unreasonable environmental laws and make tradeoffs between the environment and economic growth."

—Representative Phil Gramm of Texas

Government scientists reported in 1970 that Steubenville, Ohio, had the dirtiest air in the United States. In this steel town of 26,500 people on the west bank of the Ohio River, the prime culprit was "suspended particulates" (otherwise known as soot).

In the same year, Congress drafted the Clean Air Act—the most sweeping of the landmark antipollution laws of the 1970s. The Act set standards to protect Americans from dirty air, and established regulations requiring industry to take measures to meet those standards. Factories, mills, and power plants had to invest in costly equipment to remove some of the pollutants from the vapors they spewed into the air. Automobile makers had to retool their plants to produce gadgets that would reduce harmful emissions from tailpipes.

These efforts cost industry (and the consumer, in the form of pass-along price increases) billions of dollars, but they resulted in cleaner air, better visibility, and improved public health. Across the United States, sulfur dioxide and carbon monoxide in the air have been cut by 40 percent and suspended particulates by more than 30 percent. Between 1976 and 1981 there was a 66 percent reduction in the number of "unhealthy" days caused by high levels of carbon monoxide in eighteen of the nation's largest cities.

Even Steubenville is breathing easier these days. The soot in its air has been cut by half, and it has yielded its worst-air title to Los Angeles. Long-time Steubenville residents marvel at the (relatively) clear skies over their city, and the end of burning eyes and chronic wheezing. No longer does paint peel from their houses a week after it is applied, and they can now take long walks without coming home with a silvery complexion caused by graphite in the air.

But these improvements in Steubenville's quality of life have not come cheaply. After being fined $1.2 million for not complying with the Clean Air Act, Ohio Edison spent $450 million—$50 million more than its coal-fired power plant in the city had cost when first built—on new technology to control the plant's emissions. One solution: building taller smokestacks to disperse sulfur dioxide downwind. Unfortunately, what Steubenville lost the Northeast gained: "acid rain" caused largely by sulfur dioxide from Midwestern coal plants such as the one in that Ohio city.

In the Steubenville area, the Clean Air Act killed many jobs in coal fields, where high-sulfur coal had suddenly lost its allure. Steel plants—already hard hit by a recession and foreign competition—had to spend millions of dollars on pollution controls. In the early 1980s, the unemployment rate in the area soared past 13 percent. While only a fraction of the job loss could be directly linked to the Clean Air Act, many Ohio Valley residents perceived a cause-and-effect relationship between the two. Around Steubenville, it is often said that "Dirt is money."

When Ronald Reagan took office in 1981 he responded to these perceptions, acknowledging that "the American people insist on a quality environment," while pointing out that "we also strive for economic progress."

With the Clean Air Act due for renewal in 1982, and Congress promising to amend it so as to weaken some of its standards, the President's insistence that all regulations pass cost-benefit muster set what many environmentalists saw as a sinister tone for the debate.

Although the Clean Air Act prohibits regulatory decisions based solely on cost-benefit formulas, Administration officials said that this wouldn't stop them from carefully weighing the pluses and minuses of new proposals as part of their analysis. Representative Henry A. Waxman, chairman of the Health and Environment Subcommittee in the House, which was holding hearings on amendments to the Act, said that he suspected that the Administration would "use cost-benefit analysis to reach decisions that will favor business and industry in this country rather than the public. It will be a political tool rather than a regulatory tool."

The lines were drawn in Congress for an epic duel—perhaps

a pivotal battle in the short history of environmental protection. On one side were legislators such as Waxman, a Democrat from California, and many well-funded environmental groups lobbying for a continuation of the Clean Air Act with only minor changes; on the other, Representative John Dingell, Democrat of Michigan, and lobbyists from giant industries seeking a lightening of their regulatory load. Dingell represents Detroit while Waxman comes from the state most adversely affected by auto pollution.

The legislative battle over the costs and benefits of the Clean Air Act broke out in December 1981, when Representative Thomas A. Luken, Democrat of Ohio, introduced a bill that would serve as a vehicle for the White House and the automobile industry to make major changes in the law. The bill attracted bipartisan support. Since two of the Clean Air Act's staunchest defenders in the Senate were Republicans (Senators John Chaffee of Rhode Island and Robert Stafford of Vermont), it was apparent that the conflict would not be fought strictly along party lines.

The Luken measure—actually authored by Dingell, in the opinion of some of its opponents—called, among other things, for a doubling of the allowable limits on nitrogen oxide and carbon monoxide emissions from automobiles. It also suggested that the EPA could extend its deadlines for meeting air-quality standards by up to eight years or more. Industry spokesmen called the Luken measure a moderate bill; environmentalists said it would cut the heart out of the clean air program. *The New York Times* declared that a "showdown" in the conflict was approaching. Representative Waxman submitted his own bill, "fine-tuning" the Act (as he put it), but the environmentalists, still reeling from the smashing GOP election victories of 1980 and a series of subsequent triumphs by the Reagan forces in Congress, were clearly on the defensive.

Organized labor was divided. Building and construction trade councils lobbied Congress for revisions in the Act, claiming there were jobs to be gained by it; however, many of the AFL-CIO industrial unions opposed any revisions. One statement signed by six unions, including the United Steelworkers of America, called claims for a need to weaken the Clean Air Act so as to improve the economic situation and promote job security "simply false. . . . Pollution abatement and job security go hand in hand," the resolution continued. "Environmental regulations have not been the primary cause of even one job shutdown." And it noted that "deteriorating and obsolete facilities, which are obsolete polluters, are vulnerable to economic collapse and plant closings." John J. Sheehan, assistant to the president of the Steelworkers, pointed out one fact often overlooked in all of the economic discussions over the Clean Air Act—that workers in areas with the dirtiest air were most in need of a strong Act to preserve their health.

The lobbying on both sides was fierce. Industry lobbyists referred to the environmentalists as "greenies," "tree huggers," and "bird-watchers." Environmental groups sent protesters to Capitol Hill wearing surgical masks; some were removed from hearing rooms for distributing leaflets. The environmentalists referred to Dingell as "Tailpipe Johnny" and "dirty Dingell." Ralph Nader called Dingell "the number-one enemy for consumers on Capitol Hill."

Dingell did not deny his pro-industry bias, and understandably so. His district, which includes parts of economically hard-hit Detroit and Dearborn, is home to more than twenty steel mills, chemical factories, and auto plants, including the world ·headquarters of the Ford Motor Company. "A half million or so" of Dingell's constituents, he claims, are connected in some way to the auto industry, including his wife, who works for GM as a senior analyst in Washington. Of his four children, one works for Ford and another for an auto equipment company.

Representative Waxman called the industry/White House/Congress coalition "an unholy alliance based on greed." Common Cause revealed that ninety-three companies that had been found in violation of the Clean Air Act had donated $729,715 to the fifty-eight members of Congress who were reviewing the Act in committee. Congress Watch, a Nader organization, released a report called "The Cash Solution to Air Pollution," which examined contributions to Congressmen by the auto industry. It concluded that House members who had voted to weaken the Act in committee had received four times as much campaign money as those who had defended it.

Some of these contributions were "large enough to buy a car, and in this era of sticker shock, that's saying a lot," said Carole LeGette of Congress Watch. The sixteen House members who supported the auto industry's positions had received an average of about $5,300 from industry political-action committees (PACs).

According to Congressional aides, members of key committees were being barraged with phone calls from businessmen who claimed that they would be driven out of business if the Clean Air Act was not toned down. And some utilities enclosed scare-stories in the envelopes sent out with their monthly bills to consumers. The American Electric Power Company, which supplies power to 7 million people in the Midwest, warned, for example, that new measures to control acid rain could cost the company (and thus its customers) $2 billion a year, a figure clearly exaggerated for effect.

Cost-benefit ratios on the acid-rain issue abounded, but the Reagan Administration argued that this subject deserved much further study. "If you dig down in the glacial ice," noted Secretary of Energy James Edwards, "it's more acid than the rain we have today, so I wonder [which] smokestacks from a couple of billion years ago were responsible." (The answer: natural smokestacks—volcanoes.) Edwards also said that he didn't "want to stop acid rain, because 99.9 percent of all rain is of an acid nature," but didn't mention that the rain over much of the Northeast was twenty times more acid than usual.

The Clean Air Act debate in Congress was fascinating because at its heart it was little more than a cost-benefit analysis played out in public on a grand scale. The analysis, however, was often more emotional than technical. The environmental side pooh-poohed the costs and boosted the benefits of the Act, while industry and its allies did the opposite.

Industry's case: William D. Ruckelshaus, the once and future head of the Environmental Protection Agency, set a tone for the anti-Clean Air Act forces in a June 1981 letter to Vice President George Bush. Ruckelshaus, who had not yet left his job with the Weyerhauser Corporation to rejoin the EPA, called for major changes in the Act, including adjustments in air-quality standards. Saying that it was not possible merely to "fine tune" the law, Ruckelshaus wrote that it was "inherently impossible" to attempt to cope with air pollution by basing national policy on health or environmental effects alone. He called the Act "a complicated, pervasive law that is causing our society to spend very large sums of money for marginal benefits and thus should be changed."

Ruckelshaus cautioned that Congress was "largely ignorant of the impact of the law" and that the media was strongly biased in favor of environmental regulation. But, he affirmed: "As the first administrator of EPA, I am largely to blame for most of the bad things done to business by environmental regulation over the last decade," and added that "Given my culpability, I am mightily interested in seeing constructive change in our country's social regulatory policy."

As the clean air debate carried over into 1982, industry lobbyists focused on two themes sounded in 1981 by Ruckelshaus: that amidst a recession and with foreign competition increasing, the country could no longer afford to pay its pollution-abatement bill; and that clean air regulation had reached the point of diminishing returns.

The Chemical Manufacturers Association complained that the Clean Air Act had "spawned numerous regulations [affecting] the nation's ability to cope with other urgent necessities of our time—such as economic revitalization, energy independence, urban renewal, affordable transportation and control over inflation." And, said Jerry J. Jasinowski, chief economist of the National Association of Manufacturers, "What we are really talking about here is competition for scarce resources. . . . Historically we have biased the system toward more regulations by not making any effort to estimate the benefits, or overestimated them."

However, Jasinowski was not afraid to estimate the *costs* of the Act. He claimed that it had imposed total costs of $118 billion—or $2,100 for every four-person household in the country—over the decade. (Complying with the Act, according to other calculations, would reach costs of about $20 billion annually by the mid-1980s.) Jasinowski contended that his group was "totally committed to continued progress toward cleaner air"; what it opposed were the alleged "Catch-22" provisions of the Act, which made it difficult for a company to open a new plant in "clean" areas of the country (because the plant, even with modern pollution controls, would still dirty the air), while at the same time keeping it out of "dirty" areas (because it would make them dirtier).

As for the dwindling returns on investment in clean air, economist Lawrence White, in a study for the American Enterprise Institute, calculated that the 1979 auto-emission standards set by the Act had reduced hydrocarbon emissions by 82 percent, carbon monoxide by 83 percent, and nitrous oxides by 41 percent. Meeting these standards, he said, had added about $700 to the lifetime cost of a new car. Complying with the 1981 standards, he said, added another $700 to this cost, but resulted in smaller gains: another 13 percent cut in hydrocarbons, a 14 percent cut in carbon monoxide, and a 34 percent cut in nitrous oxides. The implication was that the first $700 chunk was worth it; the second was not.

An opinion piece supporting White's thesis appeared in The New York Times under the headline "PAYING FOR POLLUTION VIRTUE." The subhead read "CLEAN AIR'S NICE, BUT IT'S STILL RIGHT TO RECKON THE PRICE."

In doing just that, General Motors also took into account the presumed benefits of the Act's new auto emissions standards, establishing a classic cost-benefit equation. GM said that society had spent $700 million a year to reduce carbon monoxide emissions to a standard of 15 grams per mile, thus prolonging (according to its estimate) thirty thousand lives by an average of one year each. GM seemed to think that this was fair: It worked out to $23,000 for each life prolonged. But meeting the new standard of 3.4 grams per mile, said the company, would cost another $100 million, and prolong only twenty lives by one year—an estimated cost of $5 million for each new lease on life. This, in the company's view, was clearly unreasonable.

Roger B. Smith, chairman of GM, charged that "A lot of engineering dollars are spent on overkill for meeting environmental problems." Stating that relaxing the standards could mean savings of $80 to $300 per car, Smith backed the amendment then being considered by Dingell's House Energy and Commerce Committee, which would have rolled back the emissions standards from 1981 to 1980 levels. However, Smith did not—as *The New York Times* pointed out—explain how "easing pollution rules would help the industry resolve its serious problems." Nevertheless, his message was clear: What is good for General Motors is (still) good for America.

The environmentalists' case: They argued that the Clean Air Act had worked admirably and needed only a little streamlining—some red-tape cutting and the elimination of rules that needlessly delayed industrial expansion. They said that tough regulations were still necessary to protect public health and were reasonably priced, perhaps even cost-effective. And they questioned the accuracy of some of industry's cost-benefit equations. Representative Ron Wyden of Oregon referred to "voodoo environmental protection," which he defined as being able to "convince people they are breathing clean air" by "waving a magic wand and shifting some numbers around on paper."

While those arguing for a rollback in the Clean Air Act regulations often implied a declining need for tough standards, several reports indicated otherwise. One study by the Congressional Office of Technology Assessment indicated that fifty-one thousand people in North America might have died in 1980 alone from illnesses caused by sulfur pollution—nearly 2 percent of all deaths during that year in the United States and Canada. Another OTA report revealed that ozone—a form of oxygen frequently created by a reaction between nitrogen oxides and hydrocarbons in automobile exhausts (a prime cause of "smog")—was costing the nation from $2 billion to $4.5 billion annually in corn, wheat, soybean, and peanut production. And these were "conservative estimates," said Representative George E. Brown, Jr., of California, adding that the figure would be much higher if other crops and other pollutants were included. Even with matters as they were, the loss represented possibly 5 percent of the total annual farm output.

Attacking one specific proposed change in the Act, the National Audubon Society charged that 2,200 people a year in Western states alone would die needlessly if Congress deleted the rule requiring coal-burning power plants to install scrubbers in their smokestacks so as to remove sulfur dioxide. (The Reagan

Administration called the Audubon Society's warning an example of the scare tactics used by clean-air diehards.)

Maintaining the antipollution regulations would undoubtedly yield health benefits. The question that remained was: Could industry, and the American public, afford to pay the bill? In constructing their cost-benefit ratios, environmentalists had to show that money spent to keep the regulations would be well spent.

Some refused to get involved in this type of justification-by-equation, either calling it impossible to estimate how many lives were being saved and how many illnesses prevented by cleaning up the air, or stating that even if reasonable estimates were established they could not put a dollar value on these "savings." Some, however, met the dreaded cost-benefit ratio head-on.

A major study prepared for the Natural Resources Defense Council (NRDC) in Washington, for example, concluded that "the Clean Air Act has had a very minor impact on the aggregate economy and, in many instances, a positive one at that." Co-written by Adam Rose, a professor of economics at the University of California, and released in March 1982, the report stated that "Fair scrutiny of the Clean Air Act's reported economic impact fails to support the assertion that it had imposed excessive economic costs on the U.S. economy." It went on to say that while a tremendous amount of money had been spent on air-pollution abatement—$65.2 billion from 1972 to 1979—this was only a "modest" expenditure in contrast to the economy as a whole, representing only 2.7 percent of all expenditures by private industry during this period, and just 0.6 percent of the gross national product. Moreover, the report pointed out that pollution abatement expenditures had leveled off after the great surge in antipollution investment in the early 1970s.

Surprisingly, according to the report, much of this investment had not only been very good for the public—but had also been reasonably good for business, since:

- *Expenditures* by one company had represented *income* for other companies that produced machinery and materials for abatement devices. A loss of employment in one plant in one geographical area, caused by having to comply with the Clean Air Act regulations, was more than balanced by jobs created in many other (widely dispersed) companies manufacturing antipollution equipment. (The NRDC set this net job gain at 350,000–500,000.)

- Money spent on pollution-control measures was often more than regained by increased productivity. "Requirements to reduce emissions have forced firms to use scarce factors such as energy, materials, heat and capital more efficiently," said the report to the NRDC, and added that "one aspect of this effect is greater interest in recovering resources from wastes." The Westvaco Corporation, for example, found the recovery of material from mill wastes so profitable that it created a chemical subsidiary with $45 million in yearly sales, all from materials previously discarded or burned. The chairman of the Haynes Dye and Finishing Company stated that "cleaning up our stacks and neutralizing our liquids was expensive, but on balance we have actually made money on our pollution-control effort. EPA has helped our bottom line."

- Many costs to industry were simply passed along to consumers. (This is not always the case—electric utilities can pass along their costs because they are monopolies, whereas steel companies cannot because of foreign competition.)

Balanced against what the NRDC called at most a slight negative economic impact of the Clean Air Act were its many benefits. "Conventional measures of the impact of air pollution controls," said the report to the Council, "do not take into account the benefits that are not reflected in market transactions." These, explained the report, include improving health, protecting "visibility," and preventing "damage to the natural environment." A recent University of Wyoming study, for example, indicated that people living in all parts of the United States valued the clean air in the national parks and recreational areas of the Southwest at a level of $5.8 billion annually.

The most comprehensive analysis that had considered these

"intangible" factors, according to the NRDC, had been a 1979 report prepared by Bowdoin College economist A. Myrick Freeman III for the President's Council on Environmental Quality (CEQ). Using 1978 as a base, and synthesizing existing research, the study had estimated $21.4 billion in annual benefits from air-pollution control through savings from reduced damage to vegetation and crops and improvements in human health, among other factors. By comparison the CEQ had estimated the 1978 cost of compliance with federal environmental regulations at $16 billion. Commenting on the Freeman report, Robert Harris, a member of the CEQ, said: "In our desire to be economical in Federal spending and regulation we cannot be penny-wise and pound-foolish."

These charges and countercharges over the Clean Air Act— including the latest reports, ratios, and formulas—dominated activities on Capitol Hill during the spring of 1982. The Reagan Administration won some key early votes in House committees, but opposition to amending the Act grew in the Senate.

Then, on April 29, 1982, a *New York Times* headline proclaimed: "Industry Effort to Change Clean Air Act Is Stalled." The *Times* story reported that the House Energy Committee chaired by Representative Dingell had, in recent days, "unexpectedly voted . . . against a brace of changes backed by industry." The bipartisan coalition marching against the Clean Air Act was beginning to waver.

By late summer, these forces were in full retreat. In an important vote on August 11, Dingell's committee approved tough standards on thirty-seven airborne pollutants suspected of causing cancer and birth defects, beating back a more lenient, industry-supported proposal. "My strategy," Congressman Waxman later said, "was to stall until I was able to put together a majority."

This test vote signaled a dramatic change in Congressional sentiment, discouraging opponents of the Clean Air Act from further attempting to amend the bill during 1982. Supporters of the Act were pleased; the ball was still in their opponent's court (as they say in Washington). The Act would remain in effect indefinitely, without any major changes.

What was behind the remarkable turnaround in the fate of the Clean Air Act? For once both sides in the debate could agree on the reason, which had caught everyone—environmentalists, industry, the White House, the press, and Congress—by surprise. They had all misread the Reagan landslide of 1980 as a mandate to cut government regulation across the board. But what public opinion polls and constituent mail to congressmen showed instead was that a large portion of the public did not feel that clean air was negotiable; they knew what they wanted, and what they wanted was clean air at almost any cost.

Survey after survey had shown that despite broad-based support for regulatory reform in nearly every other area, support for the Clean Air Act remained strong. A Lou Harris survey revealed that 86 percent of those polled opposed weakening the Act. A *New York Times*–CBS News poll showed that most of the public supported strong environmental protection "even if it requires economic sacrifice," and that 58 percent felt that the standards for environmental protection could never be too high.

It had been this sentiment that had turned the tide in Congress in 1982. "This is an election year," commented Representative Don Albosta, a Michigan Democrat, "and there are simply more people in my district who want clean water and air and the wilderness than there are industrialists. The industrialists have more money to contribute, but the environmentalists have the people."

The environmental consensus formed in the early 1970s, and which cut across geographical and ideological lines, had held after all (and would continue to coalesce throughout 1983 and into 1984). Ronald Reagan had hoped that the desire for clean air was merely a passing fancy; apparently, it was not. EPA administrator William Ruckelshaus admitted that the Reagan team had "confused" the public's wish to improve the *way* the goals of protecting the environment and public were achieved with a desire for *changing* the goals. "We cannot deregulate in this area," Ruckelshaus said.

A Congressional aide, quoted in *The New York Times*, said: "People can see clean water and they know when they're breathing clean air. But many of them," he complained, "don't understand why they should give these up to promote productivity."

It is much more likely that the public understands all too well that clean air does not come cheaply. In fact, an August 1984 EPA report predicted that each person in the United States would in this decade pay an average of about $234 per year in increased taxes and consumer prices to maintain current air- and water-pollution-control programs. But only a tiny percentage of the American people study the facts and figures, equations and cost-benefit hoopla created by experts and promoted by opposing sides in the clean air debate; most do not even know these statistics exist. They seem to take a longer and wider view.

In one *New York Times*–CBS poll, two-thirds of the people surveyed said that they wanted a tough Clean Air Act to "maintain present environmental laws in order to preserve the environment for future generations." Often accused of being shortsighted and concerned only with bread-and-butter and their next paycheck, the American public is, in this case, apparently looking around, looking a little further down the road, and looking out for someone else.

7
The Ultimate Risk:
The Greenhouse Effect

Seventy years from now, the surface of the Earth is likely to be warmer than it has been for the past 125,000 years. It may be more than 5 degrees Fahrenheit warmer than it is today, which would approach the warmth of the Mesozoic period—the Age of the Dinosaurs. If this occurs, summer temperatures in the American Midwest will frequently reach 120 degrees Fahrenheit. Agriculture will perish in many areas. The Antarctic ice sheet may melt and the sea level around the world may rise, threatening to submerge coastal cities.

There have been warming trends before, but never one so rapid as this—virtually overnight on the geological clock. Rather than having several hundred years to cope with the changes it may bring, humankind will have to adjust in little more than half a century.

"There would be winners and there would be losers," says Dr. Robert M. White of the National Research Council, discussing the effects of a warmer Earth. "A climate change could be the cause of a major redistribution of wealth, and from the point of view of mankind, quite an arbitrary one."

More than a severe disruption of the world economy is at stake, however. The very survival of Earth's highest forms of life may be on the line. In just a couple of hundred years, the sea level may be sixty feet above today's mark. A thousand or more years down the line we face a two-hundred-fifty-foot rise—

goodbye New York, Los Angeles, and Tokyo—and a possibly unbearable tropical climate worldwide.

Now, the good news: Something can be done to prevent—or at least mitigate—this threat. On a global basis, humankind can cut down its burning of fossil fuels, stabilizing the excessive accumulation of carbon dioxide in the Earth's atmosphere that creates the hazard known as the Greenhouse Effect.

There is no sign, however, that we have the slightest interest in doing this.

Carbon dioxide in the atmosphere is thought to act like the glass of a greenhouse: It lets sunlight in, but doesn't let the heat out. Carbon dioxide absorbs heat radiated from the Earth that would otherwise dissipate into space. The more carbon dioxide in the atmosphere, acting as a kind of thermal blanket, the warmer the Earth should become, according to the Greenhouse theory, which is accepted by a vast number of (but by no means all) scientists.

Normally, carbon dioxide is not only harmless—it is absolutely essential to life. We exhale it when we breathe; trees and plants use it in the photosynthesis process, by which they produce the oxygen we breathe and crops we eat. It is found in all living things and its elements remain in things that have died. That is why the rapid worldwide surge in the burning of fossil fuels is causing such an increase in the amount of carbon dioxide in the Earth's atmosphere. "Humans are speeding up the cycle," notes climatologist David Burns of the American Association for the Advancement of Science, "by taking the carbon stored in plants over millions of years and burning it in just a century or two."

A century ago the amount of carbon dioxide in the atmosphere was 280 to 300 parts per million. It is now 335 to 340 parts per million and it is expected to reach at least 600 parts per million in the next century. This represents a doubling of atmospheric carbon dioxide within 200 years.

Carbon dioxide has been building up in the atmosphere since the Industrial Revolution ushered in the wide use of coal for fuel. But until the past three decades, when runaway industrialization caused exponential gains, the carbon dioxide buildup was steady but slow. The deforestation that has accompanied rapid industrial growth has aggravated this problem, since destroying forests allows carbon dioxide normally consumed by the trees to remain in the air.

Critics of the Greenhouse theory have pointed to an apparently important flaw in it: that during the past forty years, when the carbon dioxide content of the atmosphere has grown most rapidly, global temperatures have dipped slightly. However, these global temperature records do not take into account normal temperature cycles (which seem to cause slight warming and cooling periods about every 180 years) and other natural factors that may "mask" the growing effect of carbon dioxide. Volcanic eruptions, for example—such as that of Mexico's El Chichon volcano in the early 1980s—increase the Earth's cloud coverage, cooling its surface for months or years. Thus, supporters of the Greenhouse theory contend that even with the small temperature decline since 1940, there has been an *overall* temperature increase in the past 100 years.

Nevertheless, the lack of clear-cut evidence for a major warming effect may have tragic consequences, for it has already undermined efforts at getting governments of the world's nations to deal with the threat of such an effect—which Greenhouse theorists indicate will become indisputably apparent by the turn of the century, and will usher into the United States, a few years later, conditions resembling those of the "Dust Bowl" of the 1930s.

During that decade, an increase of just 1 degree Fahrenheit in the average temperature of the Earth's surface triggered heat waves, droughts, and tropical-storm activity unparalleled in modern times. (While this 1 degree Fahrenheit may seem small, it is an *average* of temperature highs and lows over an entire decade. One summer of destructive record heat can be offset statistically by one mildly cold winter.) In 1936 alone, it is estimated that nearly five thousand Americans lost their lives because of the heat. In that year, the average (mean) summer temperature in Kansas was 85.5 degrees Fahrenheit; the temperature topped 100 degees Fahrenheit for fifty-three days in Kansas City, Missouri; and to the north, state records of 114 degrees were set in Minnesota and Wisconsin.

If 1936-level temperatures return to America, we can anticipate (at the very least) massive crop failure, dislocation of residents, unhealthy smog and pollution caused by stagnant air masses, and unprecedented energy demands for air conditioning (which would in turn burn more fossil fuel, putting more carbon dioxide into the atmosphere). Of course, the upward "creep" in temperature would not stop there. By the year 2020 the increase could be 2 to 3 degrees. This would cause conditions analagous to those during what is known as the Medieval Warm Period (A.D. 800 to 1200), when it was so warm that wine-growing was a major industry in England.

A few years later—when many of us would still be living— the temperature increase would hit 4 to 5 degrees higher than it is today, and the climatic environment of the world would resemble the Altithermal Period of four thousand to eight thousand years ago, when camels lived in what is now Alaska, as revealed by their fossil remains.

An authoritative federal study, conducted for the National Aeronautics and Space Administration (NASA) in 1981, and based partially on computer simulations, predicted a 5 degrees Fahrenheit temperature increase within the next century if fuel burning increases at a *slow* rate (with added emphasis on alternative energy sources). If fossil fuel use grows significantly as Third World nations continue to industrialize, said the NASA report, the worldwide temperature rise could be 6 to 9 degrees. The NASA scientists also cited the observed surface temperature of Venus—900 degrees Fahrenheit—as support for the Greenhouse principle; Venus has an atmosphere formed largely of carbon dioxide.

The warming predicted by the NASA study would not be consistent in all climatic zones. It would, for example, be most extreme in the polar regions—where it would range from two to three times greater than in the tropics. Even at its lowest projected levels, however (based on a doubling of the carbon dioxide content in the air), the temperature increase would be great enough to melt away the ice in the Northwest Passage and other Arctic zones during most of the year. In NASA's report the scientists noted that "danger of rapid sea level rise is posed by the West Antarctic ice sheet which is vulnerable to rapid disintegration and melting in case of general warming." This "deglaciation" could be rapid, "requiring a century or less," according to the NASA report, and causing a rise of fifteen to twenty feet in the sea level, which "would flood 25 percent of Louisiana and Florida, 10 percent of New Jersey, and many other lowlands throughout the world."

Not all experts share this somber view of the future. At a scientific meeting held in the spring of 1984 at Columbia University, a number of specialists in polar ice caps expressed doubts that the sea level would rise as a result of the warming of the earth's climates. Scientists at the Columbia meeting sharply differed in their predictions about what would happen to the Antarctic ice sheets (whose melting could possibly raise sea levels) in the event of a substantial warming of the earth. But out of the meeting came a general consensus that too little is known at present about the factors that control the growth and shrinkage of the Antarctic ice sheets to predict what precisely would happen to them in the event of global warming.

In any event, the United States, under a siege of heat waves, would be stricken with drought. Disruption of agriculture would be widespread. Residents of California and the Southwest would use all available water for drinking; irrigated farmland would probably have to be abandoned. California's central valley farm belt would suffer so greatly that "the centers of agriculture could move back to the northeast," according to William Nierenberg of the Scripps Institute.

But the Greenhouse Effect would not be disastrous for everyone. Many of those living in temperate climates would undoubtedly suffer, but in some humid areas temperatures would rise more slowly because carbon dioxide would have to compete with moisture (water vapor) for room in the atmosphere. Such arid areas as northeastern Africa and parts of India would benefit from an increase in precipitation. Dr. Roger Revelle, a University of California professor, has projected that under these conditions the southern Sudan "could grow enough food to feed the present world population all by itself." Agriculture in parts of the Soviet Union would be hard hit, but the country's overall grain yields might increase.

For the United States, however, most experts agree with Harold Bernard, who predicted in his 1980 book, *The Greenhouse Effect*, that "the economic impact could be staggering." If the United States were to lose its capacity to export grain, Bernard wrote, about 21 percent of our current total exports would be wiped out. He also summarized the vicious cycle whereby industry might destroy the nation's economy through air pollution, pointing out that:

> the biggest single item eating away at our balance of payments now is imported oil—a fossil fuel. Our continued reliance on the fossils, then, may come back to haunt us *ad infinitum*: the fossils leading to anthropogenic climate change, which leads to drought, which leads to diminished crop yields, which lead to diminished exports, which lead to ever larger deficits in our balance of payments.

One hundred years or so from now, our climate may approach that during the end of the Mesozoic era, when a sudden warming melted Arctic ice and killed the dinosaurs—a period known as the "Time of Great Dying."

Because the Greenhouse theory is so speculative—the stuff of apocalyptic fiction (such as Arthur Herzog's novel *Heat*)—it might seem to be little more than the latest scare-story from the halls of science. In fact, it has been described and studied for well over a century.

Indeed, in 1896 Svante Arrhenius of Sweden estimated that a doubling of the atmospheric carbon dioxide content would produce a global warming that was almost in accord with today's projections. In 1938, G. S. Callender voiced concern about how human practices were influencing carbon dioxide concentrations in the air. And in the United States, the President's Science Advisory Committee highlighted the carbon dioxide problem in 1965, with particular emphasis on the Antarctic ice cap. The National Research Council concluded in 1977 that "the climatic effects of carbon dioxide release may be the primary limiting factor on energy production from fossil fuels over the next few centuries."

The predictions of the NASA study—published in *Science*

magazine in 1981 and widely reported in the American media—added fuel to this fire. It also caused a backlash. Some critics questioned the study's statistics on population and industrialization gains, or pointed out that no one knew what new forms of energy might be in use fifty years ahead, or expressed faith in the ability of the Earth's oceans to remove much of the excess carbon dioxide from the air. One skeptic, Dr. Sherwood B. Idao, a climate specialist with the Department of Agriculture, claimed that even a doubling or tripling of the atmospheric carbon dioxide content would have little effect except to *increase* agricultural productivity by 20 to 50 percent—since carbon dioxide helps plants to grow if the temperatures don't climb too high.

In 1982 the National Research Council made what it called "a second assessment" of the carbon dioxide problem in the light of criticisms of its initial probe. In this reassessment, however, it found no reason to change its earlier projections (which were much in accord with the NASA figures), and it called for comprehensive monitoring of the atmosphere and a study of the oceans for any early indication of climatic change. And noting that those who doubted the Greenhouse theory remained unconvinced, the NRC report stated that "Ultimately . . . nature will reveal to us all the truth."

Some of that grim truth was glimpsed with the release of a six-year study conducted by the Office of Naval Research, which found the polar region to be polluted "on a scale and with an intensity that could have never been imagined, even by the most pessimistic observer."

At the moment of climactic truth, how prepared will the world be? Will it be too late to prevent a true crisis?

"There is a 10 to 15 percent chance that the carbon dioxide effects will end up being trivial and a 10 to 15 percent chance [that] they will be very serious," says W. S. "Wally" Broecker, a noted expert on carbon dioxide and climate. A casual man with an engaging speaking style and a pointed sense of humor, Broecker has worked for Columbia University's Lamont-Doherty Geological Observatory since the 1950s.

"We've done a lot of work on this the past five years," Broecker says, "but the answers are not a lot better right now. We have to live with those answers, though." Because the problem "won't become critical until well into the next century," Broecker explains, society will "have to make some hard decisions on what to do about it based on a lot of ignorance, and will have to do it soon.

"There's a lot of talk about acid rain now. Acid rain is important but if you stop putting [sulfur and nitrogen oxides] into the air it will stop and lakes will flush and go back to normal. Once carbon dioxide is up there it will stay in the atmosphere until, like, the year 8000."

Broecker has attempted to interest the federal government in doing something about the carbon dioxide problem, but to little avail. "It's difficult to tamper with energy policy," he reports. "If anyone is interested in this subject in Washington, it's only for political gain. [Senator Paul] Tsongas was interested for a while, but only because he opposed the synfuels program. [Senator Charles] Percy didn't want to hear anything that might affect coal in Illinois. It would help if Congress knew what was going on. We have to tell Congress that all these things can happen, like the West Antarctic ice sheet melting in three hundred years, but you can only give them a vague idea! To really *know* anything you'll have to wait another thirty years, so we won't be able to convince Congress of anything until 2010.

"For now the government is only spending $8 million on carbon dioxide research, and the government has a way of wasting three-quarters of that. With that little money they won't even tackle a problem like the West Antarctic ice sheet. What we need is a satellite to monitor that baby constantly.

"Let's face it, we're not going to see a change in an energy policy that flies on the wings of economics. And *I'd* rather have carbon dioxide in the air than hundreds of nuclear reactors and a lot of radioactive waste. So it looks like we'll be stuck with burning fossil fuels. We're adding carbon dioxide dangerously to the atmosphere right now, without regard to the consequences. No one is paying attention. It's like we're hooked on heroin. Society is programmed into using fossil fuels and we just can't break away."

But if, as the global warming caused by the Greenhouse Effect makes itself felt, the American government does resolve to deal with the problem, these would be its options:

1. *Install "scrubbers" in smokestacks and other emission points to remove much of the carbon dioxide spewed into the air.* Because of the volume of carbon dioxide involved, the cost of this would be prohibitive. Industries in the Midwest are already fighting regulations calling for a reduction in the sulfur dioxide emissions that cause acid rain (a much more current and readily apparent problem).

2. *Develop synthetic fuels ("synfuels").* This could reduce society's dependence on fossil fuels but would, if anything, worsen the carbon dioxide build-up. Synfuels, primarily oil from shale rock and oil and gas squeezed out of coal, are even "dirtier" (in terms of the carbon dioxide they produce) than natural fossil fuels.

3. *Reduce fossil fuel use.* Fossil fuels produce at least 80 percent of America's energy (oil about 40 percent, gas 16 percent, and coal 24 percent). It is also estimated that by the year 2000, the use of oil and gas will drop while coal use will double. But while the overall reliance on fossil fuels may decrease a little, total carbon dioxide emissions will remain at least the same, since the population of the United States will increase. (And this assumes a big jump in use of nuclear and solar sources, which is far from certain.) Meanwhile, many of the less developed nations of the world will be rapidly expanding their use of fossil fuels. China, for example, will be drawing on its copious supplies of coal, more than balancing any moves toward "conservation" in the West. The Soviet Union and Canada together account for 25 percent of coal-burning, and they might not be interested in halting a trend that would improve the climate in their vast Arctic territories.

Thus, even if the United States cuts its dependence on fossil fuels, this will not have much of an impact on a worldwide basis. On this issue, action by any single nation is useless—and working out a global solution to the problem will be difficult.

But it is not likely that America will be tempted to take such a stand in any case. Clearly, our energy future—and that of much of the rest of the world—seems tied to coal, of which only about 3 percent of the world's resources have been depleted. Unfortunately, coal releases 75 percent *more* carbon dioxide than does natural gas, and 30 percent more than oil.

4. *Employ nuclear sources.* The U.S. government has tried to lead the United States down the nuclear road, but in the wake of mammoth cost overruns in the construction of new plants, chronic operating problems, and the threat of disastrous accidents at old plants, this has largely been the road not taken. Some Greenhouse theorists, however, such as author Harold Bernard, feel that the nuclear option (however unappealing) should not be ignored, particularly in view of the high health costs of using coal, which takes the lives of more than 100 miners every year, and more than $1 billion annually in federal government payments to the victims of black lung disease—to say nothing of the effects of coal-related air pollution. (By 2010 there may be as many as thirty-five thousand premature deaths annually, largely due to respiratory problems, if the burning of coal increases as expected, according to a study by scientists at the Brookhaven National Laboratory.)

"Within the near future," Bernard writes, "the choice between nuclear and fossil-generated electricity will come down to assessment of risk." Thus, he asks whether the risk of a meltdown at a nuclear plant is palatable as compared to the risks inherent in American economic dependence on imported oil, the health risks inherent in the use of coal, and the cumulative contribution of fossil-fuel burning to the Greenhouse Effect.

5. *Go solar.* Will the United States finally make the great commitment to solar energy—the only apparently safe path away from the Greenhouse Effect? If it does, this still will not be a panacea until millions of solar collectors are sold in Shanghai.

6. *Give up meat.* A vegetarian society would require much less water and cultivated land, since nearly 60 percent of all farmland goes into producing animal feed.

In the fall of 1983 it looked, for a while, as if the United States would take the lead in calling attention to (if not actually doing anything about) the coming danger. The Environmental Protection Agency warned, in a report, that the Greenhouse Effect was not only inevitable—it was imminent, with the time of arrival pegged somewhere between the years 1990 and 2000. Citing temperature rises in line with the NASA projections, EPA said that changes by the end of the twenty-first century "could be catastrophic taken in the context of today's world." By 2075, for example, the rising sea level would cost the city of Charleston, South Carolina, $1.25 billion in lost residential and industrial buildings and lost land.

"A soberness and sense of urgency should underline our response to a greenhouse warming," the report concluded. Yet little urgency was apparent in another major report, on the same subject, released just three days later by the National Academy of Sciences.

This study affirmed many of the same gloomy findings in the EPA report but its conclusions were quite different: "caution, not panic" was the order of the day. "We feel we have twenty years to examine options before we have to make drastic plans," William Nierenberg, chairman of the committee that prepared the report, said. "In that twenty years we can close critical gaps in our knowledge." Yet within twenty years the average temperature will already gone up more than one degree (according to the Academy). The Academy said it didn't believe there was a whole lot we could do to stop the Greenhouse Effect from arriving. A worldwide coal ban, the Academy said, would only delay the inevitable and was economically and politically unfeasible anyway. Instead, it urged Americans to prepare for the climatic changes; think about buying oceanfront property a little farther from the beach, for instance.

Americans who had been slapped rudely awake by the EPA report had been quickly put back to sleep. President Reagan's science adviser, George Keyworth II, called the EPA report "unnecessarily alarmist." *Time* magazine declared that "the science of the phenomenon is more interesting than frightening." George Washington University climatologist Arthur Viterito said he wondered how scientists could so confidently predict what will occur in the next 100 years when "we can't even predict the weather for tomorrow." Other scientists claimed that there was little hard evidence that the expected warming would make Antarctica melt.

Michael Oppenheimer, an atmospheric physicist, had a different view. "The Academy report," he wrote, "is skeptical about the desirability of fossil-fuel substitutes, but it presents no analysis of the cost of *not* preventing climate change. . . . Waiting years to form a policy on warming will only make the necessary adjustments harder to implement. . . . The potential cost of inaction, shunted off on future generations, is intolerable."

Several experts have recently taken exception to the notion that nothing significant can be done right now to combat the approaching Greenhouse Effect. Specialists from the Marine Biological Station at Woods Hole, Massachusetts, and the University of New Hampshire have said that reducing deforestation will dramatically slow the carbon dioxide increase. And engineers and economists at the Massachusetts Institute of Technology and Stanford University proposed that energy consumption around the world could be reduced 1 percent yearly without adversely affecting the economy.

And so, despite the warnings of dozens of scientists and writers, the world is not acting swiftly to counter the growing Greenhouse Effect. After so many warnings of environmental disaster, the public is not disinterested, but rather immune to warnings of one more "ticking time bomb" (this one barely audible, far off as it is in the upper atmosphere). And as NASA observed: "Political and economic forces affecting energy use and fuel choice make it unlikely that the carbon dioxide issue will have a major impact on energy policies until convincing observations of the global warming are in hand."

Even when the warming crunch comes, the reaction by government officials accustomed to muddling through environmental crises is unlikely to be fast or farsighted. "Crisis management" may not take place until too late.

Some observers have gone so far as to suggest that in response to a steady rise in the sea level, residents of coastal cities and low-lying regions around the world will have to take matters into their own hands and either evacuate or build dikes. The EPA has found that seawalls to protect the cities of Galveston, Texas, and Charleston would cost $800 million each. But one researcher in Washington, Kent A. Price, has said that "given the slowness of the change" in climate, "seawalls and dikes are all but certain . . . [and that] once built, it will appear cheaper to make them a bit thicker and higher than to evacuate an area." Eventually, Price notes—in a farfetched but not implausible projection—much of the human race could even find itself living below sea level, "with the probability of a catastrophic breach in the dikes growing over the centuries" and leading to the probable "repetition of the legendary sinking of Atlantis."

In that case, human adaptability—normally an admirable trait—could very well be lethal.

8
What You Can Do
About Health Risks

Although most Americans favor health promotion, few bother to promote it themselves, instead expecting doctors and the government to do more to prevent illness. Ask the average mother or father about it and they'll tell you that it's something they desire for themselves and their children. But just because they want it doesn't mean that they get it. According to the American Medical Association, there are "relatively few really health-conscious individuals," and coupled to this individual inertia is the opposition to better health care from opponents who—as we have seen—stand to gain from a more risky health environment.

Most Americans could better their health by not smoking, avoiding excessive saturated fat and cholesterol in their diets, and drinking less alcohol. Yet if every man, woman, and child in the United States did some or all of these things, the tobacco industry, the meat and dairy industries, and the liquor and beer industry would take a financial nosedive. And the pharmaceutical industry wouldn't be overjoyed either because, although they won't publicly admit it, they have a vested interest in people being sick.

Just imagine, for a moment, what would happen to every facet of this nation's medical care system if most Americans stopped smoking and quit eating saturated fats and cholesterol. Within a few years, coronary care units could be less crowded, with large numbers of respiratory therapists, cardiologists, cardiac surgeons, and cancer specialists having to take up new careers.

Television promotes health-damaging habits through the creation of physicians on TV who smoke, lead high-pressure lives, and drive recklessly. Unfortunately these TV images are based largely on fact. Go into any hospital cafeteria during lunch hour and you'll find cardiologists, an excess twenty pounds of weight bulging beneath their white coats, stuffing themselves with saturated fats and cholesterol and smoking cigarettes. Because of reports of computerized hospital machinery, daring surgery and exotic laboratory studies, the public has exaggerated expectations of what curative medicine can really do. Prevention is hardly ever dramatized, not only because it is less dramatic (and often less romantic) than curative medicine, but also because most people view it as merely a concept, without either immediate or proven long-range effects. And to embrace it means denying many of the things people like to do. In effect, by not practicing prevention, we opt for a short and sweet life at the expense of a longer, somewhat less pleasant one.

In 1983, Richard S. Schweicker, then Secretary of Health and Human Services, noted that medical science was "continuing its extraordinary progress in treating people after they get sick," but added that "the next step in health care must take us a step beyond traditional medical care to stop illness before it strikes through disease prevention and health promotion." Unfortunately, most physicians' training orients them away from the very important role they could play in this regard. They are taught to diagnose and treat diseases that can often be prevented by taking the right measures. They spend many years acquiring the knowledge and technical skill needed to deal with the consequences of risk exposure; their entire philosophical orientation is toward treating and curing. It should therefore come as no surprise that organized medicine pays little more than lip service to prevention and health promotion.

Alan Blum, a physician-writer and founding member of Doctors Ought to Care (DOC), an antismoking group, points out that in-office patient education requires time and effort for which doctors are not reimbursed. For the busy practitioner, health promotion and prevention are neither cost- nor time-effective.

Yet Blum convincingly argues that physicians could, if they wished, create an educative environment in their office waiting rooms in a number of ways. They could, for instance, offer health-education materials and magazines that portray healthful life-styles. The magazine *Runner's World*, for example, accepts no cigarette advertising, nor do some other publications, yet a far greater number of physicians subscribe to magazines that contain a dozen or more color pages of such advertising. They do so partly because publishers give them attractive subscription discounts for these magazines, but also because they don't give prevention a second thought. The National Cancer Institute survey, for example, revealed that 61 percent of the respondents said they would "very likely" follow advice from a doctor on methods to reduce cancer risks, but over 80 percent said that no physician had ever offered them such advice.

Some physicians claim that they can little influence their patients' habits and life-styles. Yet the fact that Madison Avenue has long used doctor-figure actors to promote medicines in television commercials demonstrates that people do listen to doctors. E. G. Marshall projects the image of a benevolent but authoritative doctor in commercials for an antacid. Similarly, Robert Young, long associated with the television role of Dr. Marcus Welby, sells a decaffeinated coffee in such a way as to convey the notion that it is healthful, that it reduces stress and makes people happier. Despite this, it is unlikely that most doctors will alter their attitudes toward health-care promotion unless consumer values and perceptions about it change and patients demand it.

One reason for the low public valuation of better health promotion and disease prevention is the difficulty we often have in seeing the connection between such efforts and their desirable results. This leads to doubt about the effectiveness of preventive measures. And preventive measures are often so simple that people doubt they are *capable* of warding off serious diseases whose management, on the other hand, requires complex treatment. Yet over the past several decades, the scientific evidence that health promotion and prevention do work has grown steadily, and it has often been shown that in the final analysis they result in what is, no doubt, the most dramatic of all medical accomplishments—the prolongation of life. By not smoking, using alcohol in moderation, eating three regular meals a day, maintaining a proper weight, and getting enough sleep and regular exercise, a forty-five-year-old man may live as much as eleven years longer than one who does not follow these practices, and a forty-five-year-old woman may live seven more years than her counterpart, according to a 1983 report by the Department of Health and Human Services.

Most health promotion and prevention efforts so far have focused on risk-factor intervention. There are many risk factors that affect our lives, including those discussed in this book. We now also know that a given disease, such as coronary artery disease, may have more than one risk factor, while a single risk factor, such as cigarette smoking, may cause more than one disease. Some of the most common health risks over which we have considerable control are smoking, alcohol consumption, overeating, lack of physical activity, and high blood-cholesterol and blood-sugar levels—each of which is involved, in some way, with seven of the ten leading causes of death in the United States. This means that changes in life-style and other preventive measures can appreciably reduce our risk of sickness and premature death.

The two key strategies of intervention are *primary* prevention and *secondary* prevention.

In the former, the emphasis is on identifying and quantifying certain personal habits that affect the risk of disease—such as diet, alcohol consumption, and exercise—and altering them to avoid disease. In secondary prevention, the disease process is identified (hopefully in its early stages), and measures are taken to avoid its progression. In some instances (such as in coronary atherosclerosis), the further progression of even a well-advanced disease can be halted by dietary and life-style changes.

Medical scientists have tried three techniques in primary prevention, each relying largely on an individual's willingness to change habits.

The first technique is called the "medical model." In it, an individual's risk factors are identified either in a physician's office or in a mass-screening program. Measures are then taken to reduce these risks.

The second technique consists of mass-education programs using schools, radio, television, newspapers, and magazines. In these programs, people are told about risk factors and how to avoid and reduce them. This approach has been especially successful in reducing cigarette smoking.

The third primary prevention technique involves changing an individual's product-purchasing and habits so as to reduce the risk of disease. Low-fat milk and soft margarine, for example, can be used in place of their much-higher-risk counterparts.

All three techniques, or any combination of them, have been tried in various situations over the past several decades, with remarkable success. In fact, these strategies had produced such promising results by the early 1970s that several groups of medical scientists began studies to measure their efficacy.

One such study, the Stanford Heart Disease Prevention Program, begun in 1972, focused on three risk factors associated with cardiovascular disease—cigarette smoking, elevated serum cholesterol levels, and elevated diastolic blood pressure (the diastolic pressure is the second number of the blood pressure reading).

The Stanford trial was aimed at developing and evaluating methods that would influence adults to change their living habits so as to reduce their risk of a premature heart attack or stroke. The study focused on three comparable communities in northern California, with populations between twelve thousand and fifteen thousand. Health data were gathered on a sample of persons thirty-five to fifty-nine years of age in all three communities, and the sample in one community was kept as a control against which to measure the others.

Interventional activities were launched in two of the communities, including radio programs, newspaper columns, advertising posters, billboards, and mailings of brochures. The purpose of this mass-media approach was to teach behavioral skills, inform people about risks, and change attitudes about risk-related behavior. In one of the communities, people in the highest-risk group were offered face-to-face instruction.

The results of the Stanford program were truly impressive:

- After two years, knowledge about various health risks had improved by only 6 percent in the control community, but was up 26 and 41 percent in the two communities in which interventional activities had taken place. (And up a striking 54 percent among the high-risk group that received intensive instruction.)
- The consumption of saturated fat and cholesterol fell 20 to 40 percent in the two communities in which educational programs had been run. This decline was even greater among the high-risk group, whose reported dietary behavior change was corroborated by declines in blood cholesterol levels.
- Cigarette smoking fell 7 to 24 percent in the campaign communities, whereas in the control community it fell by only 2.5 percent. The decline was a dramatic 42 percent among those persons who had received intensive instruction.
- In the control community, the risk of coronary artery disease rose by 5 percent, whereas in the campaign communities it fell by 15 to 20 percent among the general population and 30 percent among the high-risk group.

Following a careful analysis of all the data gathered during the two-year study, the project directors concluded that a community-oriented, multimedia approach could significantly increase public knowledge about risk factors and positively influence behavior that would lower the risk of heart disease and stroke.

Such unambiguous conclusions, however, have not emerged from all studies, including the Multiple Risk Factor Intervention Trial (MRFIT), set up in 1972 and 1973 by the National Heart, Lung and Blood Institute in twenty-two centers. The aim of this study was to see if efforts to reduce three risk factors for coronary heart disease—high blood cholesterol levels, a high diastolic blood pressure, and cigarette smoking—could lead to a significantly reduced death rate from coronary artery disease.

The population chosen consisted of 12,866 men between the ages of thirty-five and fifty-seven years who were considered at a high risk for dying from coronary heart disease. Half of the men were randomly assigned to a special intervention program, while the other half were left to their usual medical care.

The special intervention program consisted of a series of ten intensive group sessions that emphasized both education about the three risk factors and behavior modification. The physicians who ran these sessions advised the participants to stop smoking cigarettes, and in group discussions encouraged them to change their shopping, cooking, and eating habits. Those participants with high blood pressure were treated with medications, and their blood pressure was monitored on a regular basis. Teams of physicians, nutritionists, nurses, behavioral scientists, and general health counselors set goals for each participant and designed programs for them to achieve those goals. Participants were seen every four months—and more often if necessary—to assess changes in their risk-factor status. Members of the other group simply continued to consult their doctors as they had been doing.

In September 1982 the results of this seven-year trial were published in the *Journal of the American Medical Association*, and raised more questions than they answered. Overall, the risk factors diminished in both groups, but only to a slightly greater degree in the special intervention group than in the routine treatment group. There was no statistically significant difference in the death rate from coronary heart disease in the two groups.

One explanation for the small effect of intervention in the MRFIT—and the one that is now receiving the most attention—is that measures to reduce cigarette smoking and to lower the blood cholesterol level may have reduced the number of deaths from coronary heart disease in subgroups of the special intervention group, but subjects with high blood pressure may have had an unfavorable response to anti-hypertensive medications.

The MRFIT study shows once again that the way in which an experiment is designed cannot always take into account all the pitfalls that may occur. In this instance the control group may not have been a control group at all, but rather a group of people who were themselves reducing their own risks in various ways. And the beneficial effects of some risk-reducing measures—such as lowering the blood cholesterol level and stopping cigarette smoking—may have been offset by the adverse effects of the drugs used to lower blood pressure.

Despite the results of the MRFIT, the scientific consensus, based on several well-designed studies, is that risk intervention works to reduce disease incidence and prolong life, and a sufficient number of programs aimed at promoting risk reduction, as well as the successful efforts of millions of people employing their own efforts, confirms this.

The American Health Foundation, for example, has conducted its "Know Your Body" program among several thousand New York City school children for several years. In this program, children are screened for high serum cholesterol levels, high blood pressure, cigarette smoking, obesity, and other risk factors, and intervention is undertaken where necessary. Switzerland's National Science Foundation has successfully launched a program to reduce cardiovascular disease in communities with ten thousand to twenty thousand inhabitants.

Another interesting prevention program is being conducted in Mankato and North Mankato, Minnesota. Here a research project financed by the National Heart, Lung and Blood Institute and directed by the University of Minnesota is studying the more than thirty-seven thousand local residents to determine whether an extremely thorough community program can reduce the risk of heart attack and stroke. (The prevention program costs $400,000 annually, while American Heart Association projections place the yearly cost of cardiovascular disease in this area at about $8.7 million in medical expenses and lost wages.) At the core of the program is an effort to see whether community leaders can be more effective than physicians in promoting healthful behavior. As Broatch Saucier, the program's liaison with health professionals in the area, said: "The doctors feel impotent. They don't know how to motivate people to change their behavior."

Among the program highlights are a quit-smoking contest (first prize, a midwinter vacation at Disney World); menus in more than a dozen restaurants with small red hearts next to health-wise selections; and special shelf labels in nearly all of the area's supermarkets that single out "heart-healthy" foods (and billboards on the edges of town directing shoppers to look for the labels).

During the 1960s, an actuarial technique was developed to assess health risks. Known as Health Risk Appraisal (HRA), the technique uses information, provided by participants and related to the twelve leading causes of death, to estimate a participant's probability of dying in the next ten years. The appraised risk produced by this technique is a numerical calculation based both on risk factors for each particular participant and actuarial estimates of the mean ten-year probability of death for the participant's age/race/sex group. The questions used on Health Risk Appraisal forms cover age, race, sex, height, weight, blood pressure, cholesterol level, history of chronic bronchitis or emphysema, family health history, and such life-style issues as smoking, drinking, seat-belt usage, and exercise habits.

The HRA has become extremely popular in helping people to identify the risks associated with both their health status and their personal habits, after which they can do things to reduce these risks. It is not without pitfalls, however, and is often simplified into general charts whose predictions are erratic. For example, under the category "smoking," some popular charts list "no smoking" or "cessation for ten years" as representing no risk; half a pack a day as a substantial risk; one pack a day as a heavy risk; and two or more packs a day as a dangerous risk. While such guides may have some worth in motivating people to stop smoking, they are not based on hard scientific evidence, and their risk categories clearly do not always convey a sense of *quantitative* risk (actual mathematical data).

However, even though HRA forms are based on questionnaires whose reliability and validity have been challenged by some scientists, and although they are inadequate in assessing the complex relationships between various risks and disease—such as diet, lack of exercise, and smoking and coronary artery disease—they do at least provide some rough guidelines that can motivate us to change our habits.

Another popular technique developed to assess and reduce risks was the periodic health examination. During the 1960s such examinations were widely promoted among many groups, including corporate executives and the rank-and-file members of labor unions. The alleged benefits of these examinations were almost taken for granted without a critical review.

However, in 1975, two physicians, Dr. Paul S. Frame and Dr. Stephen J. Carlson, challenged the conventional wisdom that periodic health examinations were good for everyone. They reviewed thirty-six common diseases, the risk factors associated with them, and their progression with and without treatment. They also evaluated the feasibility of screening for each disease and examined the justifications given for screening for these diseases based on criteria that included each disease's effect on the quality of life, and the ability of tests to detect the disease in people with no symptoms. They concluded that wholesale screening did little to reduce risk, and proposed that tests and examination procedures be selected in relation to a person's age and sex.

Four years later, a Canadian Task Force criticized the routine physical examination, charging that it did not reflect the needs and risks of different age groups. It found that many of the tests used in the examination, such as chest X rays and electrocardiograms, were often not useful when done only on an annual basis. The Task Force recommended that annual checkups be replaced by selective examinations whose content and tests matched the needs of the patient.

In 1980 the American Cancer Society evaluated nine cancer screening tests and procedures, including their costs and benefits as well as their medical effectiveness. It concluded that medical examinations were more effective when related to a patient's age and sex group. Finally, in 1983 the American Medical Association suggested that healthy young adults under the age of forty should have medical evaluations only every five years, and every one to three years (depending on many factors) after the age of forty.

These studies and analyses have led many physicians and

physician groups (such as the American College of Physicians) to adopt individualized preventive-health-care plans, in which the tests and procedures are tailored to the needs, risks, age, and sex of each patient. Thus, unnecessary tests are avoided and specific ones are used, resulting in a more cost-effective approach than the general physical examination.

For the general population, the relaxation of the Automatic Annual Checkup rule is undoubtedly a good thing. But while most people may not require checkups, they will continue to be lifesavers for specific individuals. Many people, therefore, are leery of following the new guidelines.

In an article in *The New York Times* in 1982, Leonard Sloane reported that many of his friends said that he was wasting time and money by continuing the yearly checkup routine—until both he and his wife, Annette, got cancer "that was detected as a direct result of annual physical examinations." Sloane wrote: "Admittedly, yearly physicals are expensive and often turn up nothing. Yet for our own peace of mind, Annette and I had made a decision to have them regardless of expense. . . . What if one of us was in that small minority that might benefit immeasurably from the checkup? How would anyone react if he or she was the person for whom the annual examination detected something? Would the percentages matter then?" The Sloanes continue to see their doctor every year because "we like his batting average."

It is now abundantly clear that risk factors largely predict and determine the extent of chronic disease and premature death. Reckless driving, smoking, drug use, unhealthy diets, polluted air, chemical toxins, high blood pressure—and natural disasters such as earthquakes and storms—all constitute risks. Yet there is much we can do about them no matter what our age. Science has now provided us with the knowledge and techniques to reduce—if not eliminate—risks during pregnancy, birth, at the workplace, in the home, in our middle years, and during old age. Health Risk Appraisal and the preventive medical examination are but two of the tools at our disposal for reducing risks.

Our "tragic" awareness of risks often makes us feel helpless, victimized, only too cognizant of what Alfred North Whitehead called "the solemnity of the remorseless working of things." Risks are often beyond our control. But for the most part, risk *reduction* is in our own hands. When we decide to reduce risks, we in effect transform what were once considered acceptable risks into unacceptable ones.

Sources

GENERAL

Boffey, P. M., "After Years of Cancer Alarms, Progress Amid the Mistakes," *The New York Times*, March 20, 1984, p. C1.

Brody, J. E., *Jane Brody's Nutrition Book*. New York: Bantam Books, 1982.

Cohen, J., and Rogers, J., *On Democracy*. New York: Penguin Books, 1983.

Douglas, M., and Wildavsky, A., *Risk and Culture: An Essay on the Selection of Technological and Environmental Dangers*. Berkeley, Calif.: University of California Press, 1982.

Epstein, S., *The Politics of Cancer*. San Francisco: Sierra Club Books, 1978.

Faber, M. M., and Reinhart, A. M., *Promoting Health Through Risk Reduction*. New York: Macmillan Publishing Co., Inc., 1982.

Fischoff, B., Lichtenstein, S., Slovic, P., Derby, S. L., and Keeney, R. L., *Acceptable Risk*. Cambridge, Mass.: Cambridge University Press, 1981.

Lowrance, W. W., *Of Acceptable Risk: Science and the Determination of Safety*. Los Altos, Calif.: William Kaufmann, Inc., 1976.

Nelkin, D., and Brown, M. S., *Workers at Risk: Voices from the Workplace*. Chicago: University of Chicago Press, 1984.

Perrow, C., *Normal Accidents: Living with High-Risk Technologies*. New York: Basic Books, 1984.

Smith, T. B., "A Market Basket of Food Hazards," *A Report for Public Voice for Food and Health Policy*, Washington, D.C., 1983.

"Mr. Ruckelshaus as Caesar" (editorial), *The New York Times,* July 16, 1983, p. 22.

"Residents of Tacoma Disagree on Health Risks of Copper Plant," *The New York Times,* July 14, 1983, p. A12.

Chapter 1
CHOSEN RISKS

Fishbein, M., and Ajzen, I., *Belief, Attitude, Intention and Behavior.* Reading, Mass.: Addison-Wesley Publishing Co., 1975.

Hammond, K. R., and Adelman, L., "Science, Values and Human Judgment," *Science,* 194:389–396, 1976.

Jungermann, H., "Speculations About Decision—Theoretics, Aids for Personal Decision Making," *Acta Psychologica,* 45:7–34, 1980.

Kates, R. W., *Hazard and Choice Perception in Flood Plain Management* (Research Paper no. 78). Chicago: University of Chicago Press, 1962.

Keeney, R. L., and Raiffa, H., *Decisions With Multiple Objectives: Preferences and Value Tradeoffs.* New York: John Wiley & Sons, 1976.

Nelkin, D., and Brown, M. S., *Workers at Risk: Voices from the Workplace.* Chicago: University of Chicago Press, 1984.

Randal, J., "Bigger Breasts Are Not Necessarily Better," *The Washington Post,* May 15, 1983, p. B1.

Lowrance, W. W., *Of Acceptable Risk: Science and the Determination of Safety.* Los Altos, Calif.: William Kaufmann, Inc., 1976.

Skow, J., "Risking It All," *Time,* August 29, 1983, pp. 50–59.

Close-Up (1)
SMOKING

Coleman, M., "The Research Smokescreen: Moving from Academic Debate to Action on Smoking," *New York State Journal of Medicine,* 83, 13:1280–1281, 1983.

Cummins, K., "The Cigarette Makers: How They Get Away with Murder with the Press as an Accessory," *The Washington Monthly,* XVI, 3:14–23, 1984.

Blum, A. (editor), "The Cigarette Pandemic," *New York State Journal of Medicine,* 83, 13, 1983.

Blum, A., "Confronting America's Most Costly Health Problem: A Dialogue with Surgeon General Koop," *New York State Journal of Medicine,* 83, 13:1260–1263, 1983.

Boffey, P. M., "Health Groups Assail Cigarette Ads," *The New York Times,* February 17, 1984, p. 17.

Borgatta, E. F., and Evans, R., *Smoking, Health and Behavior*. Chicago: Aldine Publishing Company, 1968.

Brody, J. E., "The Growing Militancy of the Nation's Nonsmokers," *The New York Times*, January 15, 1984, p. E6.

Ebert, R. V., "Coal Smoke and Cigarette Smoke," *New England Journal of Medicine*, 304, 24:1486–1487, 1981.

Godber, G., "Health Versus Greed," *New York State Journal of Medicine*, 83, 13:1248–1249, 1983.

Greenberg, R. A., Haley, N. J., Etzel, R. A., and Loda, F. A., "Measuring the Exposure of Infants to Tobacco Smoke," *New England Journal of Medicine*, 310, 17:1075–1078, 1984.

Hartz, A. J., Barboriak, P. N., Anderson, A. J., et al., "Smoking, Coronary Artery Occlusion, and Nonfatal Myocardial Infarction," *Journal of the American Medical Association*, 246, 8:851–853, 1981.

Hinds, M. de C., "Senators Endorse Tougher Cigarette Warning," *The New York Times*, June 23, 1983, pp. A1, A19.

Jaynes, G., "Tobacco Industry Faces Change and Challenge," *The New York Times*, September 6, 1982, p. A1.

Kannel, W. B., "Cigarettes, Coronary Occlusions, and Myocardial Infarction" (editorial), *Journal of the American Medical Association*, 246, 8:871–872, 1981.

Krasnegor, N. A., *The Behavioral Aspects of Smoking*, National Institute on Drug Abuse Monograph 26, Rockville, Maryland, U.S. Department of Health, Education and Welfare, 1979.

Luce, B. R., and Schweitzer, S. O., "Smoking and Alcohol Abuse: A Comparison of Their Economic Consequences," *New England Journal of Medicine*, 298:569–570, 1978.

Pertschuk, M., "The Politics of Tobacco: Curse and Cure," *New York State Journal of Medicine*, 83, 13:1275–1277, 1983.

Ravenholt, R. T., "Radioactivity in Cigarette Smoke" (letter), *New England Journal of Medicine*, 307, 5:312, 1982.

Reinhold, R., "Surgeon General Report Broadens List of Cancers Linked to Smoking," *The New York Times*, February 23, 1982, pp. A1, C2.

Rickert, W. S., "Less Hazardous Cigarettes: Fact or Fiction," *New York State Journal of Medicine*, 83, 13:1269–1272, 1983.

Shapiro, H. D., "Quit Smoking, and Cut Insurance Fees," *The New York Times*, September 4, 1983, p. F9.

Whitehead, T., *Selling Death: Cigarette Advertising and Public Health*. New York: Liveright, 1975.

Wynder, E. L., and Hoffman, D., "Tobacco and Health: A Societal Challenge," *New England Journal of Medicine*, 300:894, 1979.

American Medical Association—Education and Research Foundation Committee for Research on Tobacco and Health, *Tobacco and Health*, Chicago, Illinois, AMA—ERF, 1978.

Calling It Quits: The Latest Advice on How To Give Up Cigarettes, Bethesda, Maryland, National Cancer Institute, U.S. Department of Health, Education and Welfare, 1979.

"Can We Have An Open Debate About Smoking?" (advertisement), R. J. Reynolds Tobacco Company, *Newsweek*, February 6, 1984.

Controlling the Smoking Epidemic, Technical Report Series no. 636, Geneva, World Health Organization, 1979.

The Health Consequences of Smoking, Washington, D.C., U.S. Public Health Service, U.S. Department of Health, Education and Welfare, Public Health Publication no. 1696, 1967, revised January 1968.

The Health Consequences of Smoking—Cancer: A Report of the Surgeon General, Rockville, Maryland, U.S. Department of Health and Human Services, Office of Smoking and Health, 1982.

The Health Consequences of Smoking for Women: A Report of the Surgeon General, Rockville, Maryland, U.S. Department of Health and Human Services, 1980.

"Research Casts Doubt on Danger of Cancer to Wives of Smokers," *The New York Times*, June 20, 1981, p. 1.

"Senate Passes Tobacco Aid Curbs," *The New York Times*, July 15, 1983, p. B13.

"Smoking and Cancer," *Morbidity and Mortality Weekly Report*, 31, 7:77–80, February 26, 1982.

"Smoking and Cardiovascular Disease," *Morbidity and Mortality Weekly Report*, 32, 52:677–679, January 6, 1984.

Smoking and Health: The Health Consequences of Smoking: Cardiovascular Disease. A Report of the Surgeon General, Rockville, Maryland, U.S. Department of Health and Human Services, 1983.

Smoking and Health (published bimonthly), Washington, D.C., U.S. Department of Health and Human Services, Office on Smoking and Health.

Smoking and Health: Report of the Advisory Committee to the Surgeon General of the Public Health Service, Washington, D.C., U.S. Public Health Service, U.S. Department of Health, Education and Welfare, Public Health Service Publication no. 1103, 1964.

Smoking and Health: A Report of the Surgeon General of the U.S. Public Health Service, Washington, D.C., U.S. Department of Health, Education and Welfare, 1979.

Smoking and Health: Summary and Report of The Royal College of Physicians of London on Smoking in Relation to Cancer of the Lung and Other Diseases. London: Pitman Publishing Corporation, 1962.

Smoking and Health: Report of The Royal College of Physicians of London. London: Pitman Publishing Corporation, 1977.

Smoking and Its Effects on Health, Report of a WHO Expert Committee,

Technical Report Series no. 568, Geneva, World Health Organization, 1975.

State Legislation On Smoking and Health 1980, Atlanta, Georgia, Centers for Disease Control, U.S. Department of Health and Human Services, 1981.

"Study Finds New Threat to Smokers," *San Francisco Chronicle*, March 18, 1982, p. 26.

"U.S. Health Officials Endorse Stronger Cigarette Warnings," *The New York Times*, March 12, 1983, p. A12.

Why Do You Smoke? Bethesda, Maryland, National Cancer Institute, U.S. Department of Health, Education and Welfare, 1978.

1980 Bibliography On Smoking and Health, Rockville, Maryland, U.S. Department of Health and Human Services, Office on Smoking and Health, 1980.

1980 Directory of On-Going Research in Smoking and Health, Rockville, Maryland, U.S. Department of Health and Human Services, Office on Smoking and Health, 1980.

"17,000 Fatalities Linked To Smoking: U.S. Says Heart Disease Tied to Cigarettes Will Kill 10% of the Population Early," *The New York Times*, November 18, 1983, p. D18.

Close-Up (2)
SEAT BELTS

Brody, J. E., "Seat Belts Can Lessen Injuries," *The New York Times*, December 9, 1981.

Gallup, G., "Sharp Decline Found in Seat Belt Use," *The Gallup Poll*, September 5, 1982.

Karr, A., "Saga of the Air Bag," *The Wall Street Journal*, November 11, 1976.

Lund, A., and Williams, A., "Public Opinion Surveys on Occupant Crash Protection," Insurance Institute for Highway Safety, Research Note no. 102, July 1982.

Robertson, L. S., Kelley, A. B., O'Neill, B., et al., "A Controlled Study of the Effect of Television Messages on Safety Belt Use," *American Journal of Public Health*, 64, 11:1071–1080, 1974.

Motor Vehicle Safety, 1979, Washington, D.C., National Highway Traffic Safety Administration, 1979.

Promotional Campaigns to Increase Seat Belt Use: A Review, Washington, D.C., Insurance Institute for Highway Safety, July 1981.

Close-Up (3)
EARTHQUAKES

Nigg, J., et al., *Earthquake Threat*, Report for Institute for Social Science Research, Los Angeles, California, 1979.

Penick, J., *The New Madrid Earthquakes*. Columbia, Mo.: University of Missouri Press, 1981.

Rogers, M., "When the Earthquake Hits Los Angeles," *Rolling Stone*, October 28, 1982, p. 150.

Turner, R., et al., *Community Response to Earthquake Threat In Southern California*, A Report for the Institute for Social Science Research, Los Angeles, California, 1981.

Wellborn, S., "Is a Killer Earthquake Due To Hit The U.S.?", *U.S. News and World Report*, November 14, 1983, pp. 74–76.

California Earthquake Response Plan, Southern San Andreas Fault, Sacramento, California, California Office of Emergency Services, July 1983.

State of California Seismic Safety Commission Earthquake Preparedness Task Force Report for 1982–83, Sacramento, California, June 30, 1983.

Chapter 2
THE "INFORMED" PUBLIC

Arnold, D. L., Moodie, C. A., Grice, H. C., et al., *Long-Term Toxicity of Orthotoluene—Sulfonamide and Sodium Saccharin in the Rat*, An Interim Report, Ottawa, Canada, Health Promotion Branch, National Health and Welfare Ministry, 1977.

Batzinger, R. P., Ou, S. L., and Bueding, E., "Saccharin and Other Sweeteners: Muta-genic Properties," *Science*, 198:944–946, 1977.

Boffey, P. M., "The Debate Over Dioxin," *The New York Times*, June 25, 1983, p. 1.

Cohen, B. L., "Relative Risks of Saccharin and Caloric Ingestion," *Science*, 199:983, 1978.

Feingold, B., *Why Your Child Is Hyperactive*. New York: Random House, 1975.

Hunter, B. T., *Food Additives and Federal Policy: The Mirage of Safety*. New York: Charles Scribner's Sons, 1975.

Reinhold, R., "Many Questions but Few Answers For People of Dioxin-Periled Town," *The New York Times*, December 30, 1982, pp. A1, A8.

Reinhold, R., "Missouri Now Fears 100 Sites Could Be Tainted By Dioxin," *The New York Times*, January 18, 1983, pp. A1, A23.

Rutstein, D. D., "Controlling the Communicable and the Man-Made Diseases," *New England Journal of Medicine*, 304, 23:1422–1424, 1981.

Sheppard, N., "In Dioxin-Tainted Town, No 'Welcome' Signs," *The New York Times*, January 10, 1983, pp. A1, A14.

Silverman, M., and Lee, P. R., *Pills, Profits and Politics*. Berkeley, Calif.: University of California Press, 1974.

Sipher, A. T., "The GRAS List Review," *FDA Papers*, Washington, D.C., U.S. Department of Health, Education and Welfare, December 1970–January 1971.

Slovic, P., "The Psychology of Protective Behavior," *Journal of Safety Research*, Summer 1978, pp. 58–68.

Turner, J. S., *The Chemical Feast*. New York: Grossman, 1970.

Young, J. H., "Saccharin: A Bitter Regulatory Controversy," in *Research in the Administration and Public Policy*, edited by F. B. Evans and H. T. Pinkett. Cambridge, Mass.: Harvard University Press, 1975.

The Calorie Control Council's Submission to the National Academy of Sciences Committee for a Study on Saccharin and Food Safety Policy, public meeting on saccharin, National Academy of Sciences, Washington, D.C., June 19, 1978, Calorie Control Council, Atlanta, Georgia, 1978.

Evaluation of Certain Food Additives, Technical Report Series no. 669, Geneva, World Health Organization, 1981.

Federal Food, Drug, and Cosmetic Act, As Amended, Washington, D.C., U.S. Department of Health, Education and Welfare, Food and Drug Administration, 1969.

"Low Amounts of Saccharin Cause Health Problems in Rats," *News Tribune* (New Jersey), January 4, 1983, p. 2.

The National Advisory Committee on Hyperkinesis and Food Additives. Final Report to the Nutrition Foundation, Washington, D.C., The Nutrition Foundation, 1980.

"That Pesky Food Additives Law," *Business Week*, March 12, 1960, p. 113.

Saccharin Study and Labeling Act, 95th Congress House Conference Report no. 95–810, Washington, D.C., U.S. Government Printing Office, 1977.

Saccharin: Technical Assessment of Risks and Benefits, Report No. 1, Washington, D.C., Committee for a Study on Saccharin and Food Safety Policy, National Research Council, National Academy of Sciences, 1978.

Close-Up
SATURATED FATS AND CHOLESTEROL

Altman, L. K., "Report About Cholesterol Draws Agreement and Dissent," *The New York Times*, May 28, 1980, p. A16.

Anitschkow, W., "Experimental Arteriosclerosis in Animals," in *Arteriosclerosis*, edited by E. V. Cowdry. New York: Macmillan Publishing Co., Inc., 1933, p. 271.

Bishop, J., "Scientists Are Firming Link to Cholesterol with Coronary Disease," *The Wall Street Journal*, January 10, 1984, p. 1.

Blackburn, H., *Preventive Cardiology in Practice: Minnesota Studies on Risk Factor Reduction*, Proceedings of the Conference on Heart Disease and Rehabilitation: State of the Art 1978, Milwaukee, Wisconsin, May 11–13, 1978; Boston: Houghton, Mifflin, 1979.

Boffey, P. M., "Study Backs Cutting Cholesterol to Curb Heart Disease Risk," *The New York Times*, January 13, 1984, pp. A1, A10.

Brody, J. E., "Diet to Prevent Heart Attacks Aims to Cut Blood Fat Levels," *The New York Times*, May 16, 1984, p. A16.

Brody, J. E. "Panel Reports Healthy Americans Need Not Cut Intake of Cholesterol: Nutrition Board Challenges Notion That Such Dietary Change Could Prevent Heart Disease," *The New York Times*, May 28, 1980, pp. A1, A16.

Brody, J. E., "Whole Cities Organize to Fight Heart Disease," *The New York Times*, September 6, 1983, pp. C1, C6.

Brown, M. S., and Goldstein, J. L., "Lowering Plasma Cholesterol by Raising LDL Receptors," *New England Journal of Medicine*, 305, 9:515–517, 1981.

Burros, M., "Controversial Diet to Be Published," *The New York Times*, August 24, 1982, p. 24.

Christopher, T. W., and Dunn, C. W., *Special Federal Food and Drug Laws: Statistics, Regulations, Legislative History and Annotations*. Chicago: Commerce Clearinghouse, 1954.

Farquhar, J. W., *The American Way of Life Need Not Be Hazardous to Your Health*. New York: W. W. Norton, 1979.

Gordon, T., and Kannel, W. B., "The Effects of Overweight on Cardiovascular Disease," *Geriatrics*, 28:80–88, 1973.

Hausman, P., *Jack Sprat's Legacy: The Science and Politics of Fat and Cholesterol*. New York: Richard Marek Publishers, 1981.

Hill, M. M., and Cleveland, L. E., "Food Guides: Their Development and Use," *Nutrition Program News*, Washington, D.C., Consumer Food Economics Institute, U.S. Department of Agriculture, July/October, 1970.

Keys, A. (editor), *Coronary Heart Disease in Seven Countries*, American Heart Association Monograph Number 29, *Circulation*, 41, Supplement I, 211 pp., 1970.

Keys, A. (editor), *Seven Countries—A Multivariate Analysis of Death and Coronary Heart Disease in Ten Years*. Cambridge, Mass.: Harvard University Press, 1980.

King, S. S., "Plan to Curb Dairy Surplus Goes Into Effect," *The New York Times*, December 2, 1982, p. A27.

McCollum, E. V., *From Kansas Farm Boy to Scientist, The Autobiography of Elmer Verner McCollum*. Lawrence, Kansas: University of Kansas Press, 1964.

McGill, H. (editor), *The Geographic Pathology of Atherosclerosis*. Baltimore: Williams and Wilkins, 1968.

Marmot, M. G., Syme, S. L., Kagan, A., et al., "Epidemiologic Studies of Coronary Heart Disease and Stroke in Japanese Men Living in Japan, Hawaii and California: Prevalence of Coronary and Hypertensive Heart Disease and Associated Risk Factors," *American Journal of Epidemiology*, 102:511–525, 1975.

Mayer, J., *U.S. Nutrition Policies in the Seventies*. San Francisco: Freeman and Co., 1973.

Michel, R., Quote in "Report of the Chairman of the Board," *Common Cause*, August 1982, p. 12.

Shekelle, R. B., Shryock, A. Mac., Paul, O., et al., "Diet, Serum Cholesterol, and Death from Coronary Artery Disease: The Western Electric Study," *New England Journal of Medicine*, 304, 2:65–70, 1981.

Stamler, J., "Lifestyles, Major Risk Factors, Proof and Public Policy," *Circulation*, 58, 1:3–19, 1978.

Walles, C., "Hold the Eggs and Butter: Cholesterol Is Proved Deadly, and Our Diet May Never Be the Same," *Time*, 123, 13:36–38, 40–43, March 26, 1984.

Willett, W. C., and MacMahon, B., "Diet and Cancer—An Overview"—first of two parts, *New England Journal of Medicine*, 310, 10:633–638, 1984.

Willett, W. C., and MacMahon, B., "Diet and Cancer—An Overview"—second of two parts, *New England Journal of Medicine*, 310, 11: 697–703, 1984.

"American Heart Association Special Report. Recommendations for Treatment of Hyperlipidemia in Adults," A Joint Statement of the Nutrition Committee and the Council on Atherosclerosis, AHA, *Circulation*, 69, 5:1067A–1090A, 1984.

Counseling the Patient with Hyperlipidemia, Dallas, Texas, American Heart Association, 1984.

"Dairy Farmers to Begin Paying Fee on Sales," *The New York Times*, November 28, 1982, p. 34.

Diet, Nutrition and Cancer, Washington, D.C., Committee on Diet, Nutrition and Cancer, National Research Council, National Academy Press, 1982.

Dietary Goals for the United States, Washington, D.C., U.S. Senate Select Committee on Nutrition and Human Needs, 1977.

Eating for a Healthy Heart, Dallas, Texas, American Heart Association, 1984.

"Eating the Moderate Fat and Cholesterol Way" (deleted chapter), *Food 2*, Washington, D.C., U.S. Department of Agriculture, 1982.

Federal and State Standards for the Composition of Milk Products, Washington, D.C., U.S. Department of Agriculture, 1980.

"The FTC and Cholesterol," *The New York Times*, December 15, 1982, p. A26.

Healthy People: The Surgeon General's Report on Health Promotion and Disease Prevention, Washington, D.C., U.S. Department of Health, Education and Welfare, 1979.

"The Lipid Research Clinics Coronary Primary Prevention Results: I. Reduction in Incidence of Coronary Heart Disease," *Journal of the American Medical Association*, 251, 3:351–465, 1984.

"The Lipid Research Clinics Coronary Primary Prevention Trial Results: II. The Relationship of Reduction in Incidence of Coronary Heart Disease to Cholesterol Lowering," *Journal of the American Medical Association*, 251, 3:365–374, 1984.

"Multiple Risk Factor Intervention Trial, Risk Factor Changes, and Mortality Results," *Journal of the American Medical Association*, 248, 12:1465–1467, 1982.

Nutrition and Your Health: Dietary Guidelines for Americans, Washington, D.C., Department of Agriculture and U.S. Department of Health and Human Services, 1980.

Prevention, '82, Washington, D.C., U.S. Department of Health and Human Services, 1982.

"Report of the Chairman of the Board," *Common Cause*, August 1982, pp. 11–12.

U.S. Senate Committee on Agriculture, Nutrition and Forestry, Hearing: Heart Disease: Public Enemy No. 1, May 22, 1979.

U.S. Senate Select Committee on Nutrition and Human Needs, Hearing: Diet Related to Killer Diseases, volume 2, part I, Cardiovascular Disease, February 1–2, 1976.

U.S. Senate Select Committee on Nutrition and Human Needs, Hearing: Diet Related to Killer Diseases, volume 6, Response Regarding Eggs, July 26, 1977.

Chapter 3
IMPOSED RISKS

Atkins, R., and Linde, S. A., *Dr. Atkins' Superenergy Diet*. New York: Bantam Books, 1978.

Blum, A., "Phenylpropanolamine: An Over-the-Counter Amphetamine?" (editorial), *Journal of the American Medical Association*, 245, 3:1346–1347, 1981.

Burros, M., "Residue of Chemicals in Meat Leads to Debate on Hazards," *The New York Times*, March 15, 1983, p. 1.

Fredman, S., and Burger, R. E., *Forbidden Cures*. New York: Stein and Day, 1976.

Goldman, D., Eulkerson, C., Dixon, R., et al., *Nationwide Epidemic of Septicemia Associated with Intravenous Fluid Therapy: An Analysis Based On the CDC National Nosocomial Infections Study*, Atlanta, Georgia, U.S. Department of Health, Education and Welfare, Publication No. (HSM) 73–8181, 1972.

Greider, W., "Fines Aren't Enough," *Rolling Stone*, March 29, 1984, p. 9.

Gruson, L., "A Controversy Over Widely Sold Diet Pills," *The New York Times*, February 14, 1982, p. B12.

Hinds, M. de C., "Assessing Effects of Chemically Treated Food," *The New York Times*, March 31, 1982, pp. C1, C20.

Hinds, M. de C., "Food and Drug Agency Says Some Risk Must Be Accepted," *The New York Times*, July 25, 1982, p. E8.

Lowrance, W. W., *Of Acceptable Risk: Science and the Determination of Safety*. Los Altos, Calif.: William Kaufmann, Inc., 1976.

Mackel, D. C., Maki, D. G., Anderson, R. R., et al., "Nationwide Epidemic of Septicemia, Caused by Contaminated Intravenous Products. III. Mechanisms of Intrinsic Contamination," *Journal of Clinical Microbiology* 2, 6:486–497, 1975.

Maki, D. G., Rhame, F. S., Mackel, D. C., et al., "Nationwide Epidemic of Septicemia Caused by Contaminated Intravenous Products. I. Epidemiologic and Clinical Features," *American Journal of Medicine* 60, 4:471–485, 1976.

Mazel, J., *The Beverly Hills Diet*. New York: Macmillan Publishing Co., Inc., 1981.

Nader, R., Brownstein, R., and Richard, J., *Who's Poisoning America: Corporate Polluters and Their Victims in the Chemical Age*. San Francisco: Sierra Club Books, 1981.

Pollack, A., "Trust in Computers Raising Risk of Errors and Sabotage," *The New York Times*, August 22, 1983, p. A1.

Puzo, D. P., "Food Safety Is Doubted in Survey: Must America Fear Additives," *Los Angeles Times*, May 20, 1982, pp. 1, 22.

Silverman, M., *The Drugging of the Americas*, Berkeley, Calif.: University of California Press, 1976. ·

Silverman, M., and Lee, P. R., *Pills, Profits and Politics*. Berkeley, Calif.: University of California Press, 1974.

Silverman, M., Lee, P. R., and Lydecker, M., *Prescriptions For Death: The Drugging of the Third World*. Berkeley, Calif.: University of California Press, 1982.

"Follow Up On Septicemia Associated with Contamination of Intravenous Fluids," *Morbidity and Mortality Weekly Report*, 22:124, 1973.

Close-Up (1)
ORAFLEX

Altman, L. K., "Oraflex Promotion Produced Sales, But Not Without Side Effects," *The New York Times*, August 17, 1982, p. C4.

Deitch, R., "Suspension of Benoxaprofen (Opren)," *The Lancet*, 8294, 394–395, 1982.

Hinds, M. de C., "Arthritis Drug Ban: Pro, Con and Undecided," *The New York Times*, July 25, 1982, p. 8E.

Hinds, M. de C., "Consumer Group Seeks Ban on Arthritis Drug," *The New York Times*, July 22, 1982, p. A3.

Hinds, M. de C., "FDA Says It Overlooked Drug Data," *The New York Times*, August 3, 1982, pp. C1, C7.

Hinds, M. de C., "Report Questions Lilly Role on Drug. FDA Investigations Contend Company Knew About 26 Deaths Tied to Oraflex," *The New York Times*, December 3, 1982, p. 8.

Hinds, M. de C., "Speeding FDA Drug Review," *The New York Times*, September 22, 1982, p. C1.

Lueck, T. J., "At Lilly, the Side-Effects of Oraflex," *The New York Times*, August 15, 1982, pp. F1, F6, F7.

Lueck, T. J., "For the F.D.A., a Threat to New Procedures: The Agency Is Trying to Cut Paperwork to Speed Approval of New Drugs," *The New York Times*, August 15, 1982, p. F7.

"First Oraflex Case Decided," *The New York Times*, November 27, 1983, p. E16.

Mailgram to Physicians On Suspension of Distribution and Sales of Oraflex, Indianapolis, Eli Lilly Company, August 6, 1982.

"Sale of Arthritis Drug Suspended," *The New York Times*, August 5, 1982, p. B13.

"$6 Million Award Made in Lilly Suit: Arthritis Drug Oraflex Blamed by a Federal Jury in Death of Woman in Georgia," *The New York Times*, November 22, 1983, p. A24.

Close-Up (2)
IRRADIATION

Black, E., and Libby, L., "Commercial Food Irradiation," *The Bulletin of Atomic Scientists*, June/July 1983, pp. 48–50.

Hinds, M. de C., "Assessing Effects of Chemically Treated Foods," *The New York Times*, April 5, 1982, p. C1.

Lampe, O., "Food That Stays Fresh Forever," *Next*, November/December 1980, pp. 75–78.

Schell, D., *Modern Meat: Antibiotics, Hormones and the Pharmaceutical Farm.* New York: Random House, 1984.

Solomon, S., "How Safe Is X-Rayed Food?", *American Health*, March/April 1984, pp. 92–100.

Sterba, J., "Irradiated Food: Promise and Controversy," *The New York Times*, June 30, 1982, p. C1.

Thompson, R., Purifying Food Via Irradiation," *FDA Consumer*, October 1981, pp. 25–27.

"Food Irradiation: Ready for a Comeback," *Food Engineering*, April 1982, pp. 71–80.

Radiation Preservation of Foods: A Scientific Summary, Chicago, The Institute of Food Technologists, 1983.

Chapter 4
DEFINING RISKS

Carter, L. J., "How to Assess Cancer Risks," *Science*, 204:811–816, 1979.

Colton, J., *Statistics in Medicine.* Boston, Mass.: Little, Brown and Co., 1974.

Douglas, M., and Wildavsky, A., *Risk and Culture: An Essay On the Selection of Technological and Environmental Dangers.* Berkeley, Calif.: University of California Press, 1982.

Dowie, M., et al., "The Illusion of Safety," *Mother Jones*, June 1982, pp. 36–49.

Fischoff, B., Lichtenstein, S., Slovic, P., Derby, S. L., and Keeney, R. L., *Acceptable Risk.* Cambridge, Mass.: Cambridge University Press, 1981.

Fletcher, R. H., Fletcher, S. W., and Wagner, E. H., *Clinical Epidemiology: The Essentials.* Baltimore: Williams and Wilkins, 1982.

Kaplan, M. F., and Schwartz, S. (editors), *Human Judgment and Decision Processes.* New York: Academic Press, 1975.

Kastenberg, W., McKone, T., and Okrent, D., *On Risk Assessment in the Absence of Complete Data* (UCLA Report no. ENG-7677). Los Angeles: University of California Press, 1976.

Lowrance, W. W., *Of Acceptable Risk: Science and the Determination of Safety.* Los Altos, Calif.: William Kaufmann, Inc., 1976.

Moroney, M. J., *Facts from Figures.* London: Penguin Books, 1951.

Mosteller, F., Rourke, R. E. K., and Thomas, G. B., *Probability and Statistics.* Reading, Mass.: Addison-Wesley Publishing Co., 1961.

Perrow, C., *Normal Accidents: Living With High Risk Technologies.* New York: Basic Books, 1984.

Riegelman, R. K., *Studying a Study and Testing a Test: How to Read the Medical Literature*. Boston, Mass.: Little, Brown and Co., 1981.

Roht, L. H., Selwyn, B. J., Holguin, J. H., and Christensen, B. L., *Principles of Epidemiology: A Self-Teaching Guide*. New York: Academic Press, 1982.

Rothman, K. J., "The Rise and Fall of Epidemiology, 1940–2000 A.D.," *New England Journal of Medicine*, 304, 10:600–602, 1981.

Sackett, D. L., and Gent, M., "Controversy in Counting and Attributing Events in Clinical Trials," *New England Journal of Medicine*, 301, 26:1410–1412, 1979.

Shribman, D., "Calculating the Odds on Accurate Risk Assessment," *The New York Times*, January 2, 1983, p. E6.

"3 Ex-Officials of Major Laboratory Convicted of Falsifying Drug Tests," *The New York Times*, October 22, 1983, pp. 1, 13.

Close-Up (1)
THE PILL

Altman, L. K., "Health Benefits of the Pill Found to Outweigh Its Drawbacks," *The New York Times*, July 13, 1982, pp. C1–C2.

Boffey, P. M., "Injected Contraceptive: Hazard or a Boon?", *The New York Times*, January 11, 1983, pp. C1, C3.

Cramer, D. W., Hutchison, G. B., Welch, W. B., et al., "Factors Affecting the Association of Oral Contraceptives and Ovarian Cancer," *New England Journal of Medicine*, 307, 17:1047–1050, 1982.

Inman, W. H. W., Vessey, M. P., Westerholm, B., et al., "Thromboembolic Disease and the Steroidal Content of Oral Contraceptives: A Report to the Committee on the Safety of Drugs," *British Medical Journal*, 2:203–209, 1970.

Jordan, W. M., "Pulmonary Embolism," *Lancet*, 2:1146–1147, 1961.

Ory, H. W., Rubin, G. L., Jones, V., et al., "Mortality Among Young Black Women Using Contraceptives," *Journal of the American Medical Association*, 251, 8:1044–1048, 1984.

Petitti, D. B., and Wingard, J., "Use of Oral Contraceptives, Cigarette Smoking and Risk of Subarachnoid Hemorrhage," *Lancet*, 2:234–236, 1978.

Pike, M. C., Henderson, B. E., Krailo, M. D., et al., "Breast Cancer in Young Women and Use of Oral Contraceptives: Possible Modifying Effect of Formulation and Age at Use," *Lancet*, 8375, 926–930, 1983.

Ramcharan, S. (editor), *The Walnut Creek Contraceptive Drug Study: A Prospective Study of the Side Effects of Oral Contraceptives*, Washington, D.C., U.S. Department of Health, Education and Welfare, volume 1 (1974), volume 2 (1976), volume 3 (1981).

Sartwell, P. E., Masi, A. T., Arthes, F. G., et al., "Thromboembolism and Oral Contraceptives: An Epidemiologic Case Control Study," *American Journal of Epidemiology*, 90:365–380, 1969.

Stadel, B. V., "Oral Contraceptives and Cardiovascular Disease"—first of two parts, *New England Journal of Medicine*, 305, 11:612–618, 1981.

Stadel, B. V., "Oral Contraceptives and Cardiovascular Disease"—second of two parts, *New England Journal of Medicine*, 305, 12:672–677, 1981.

Stone, D., Shapiro, S., Kaufman, D. W., et al., "Risk of Myocardial Infarction in Relation to Current and Discontinued Use of Oral Contraceptives, *New England Journal of Medicine*, 305, 8:420–424, 1981.

Vessey, M. P., "Female Hormones and Vascular Disease: Epidemiologic Overview, *British Journal of Family Planning*, 6: Supplement:1–12, 1980.

Vessey, M. P., "Steroid Contraception, Venous Thromboembolism and Stroke: Data from Countries Other than the U.S.," in *Risk Benefits and Controversies in Fertility Control*, edited by J. J. Sciarra, G. I. Zatuchini and J. J. Speidel. Hagerstown, Md.: Harper and Row, 1978, pp. 113–1111.

Vessey, M. P., and Doll, R., "Investigation of Deaths from Pulmonary Coronary and Cerebral Thrombosis and Embolism in Women of Childbearing Age," *British Medical Journal*, 2:193–199, 1968.

Vessey, M. P., Lawless, M., McPherson, K., et al., "Neoplasm of the Cervix Uteri and Contraception: A Possible Adverse Effect of the Pill," *Lancet*, 8375, 930–935, 1982.

Vessey, M. P., McPherson, K., and Johnson, B., "Mortality Among Women Participating in the Oxford Family Planning Association Contraceptive Study," *Lancet*, 2:731–733, 1977.

"Collaborative Group for the Study of Stroke in Young Women. Oral Contraception and Increased Risk of Cerebral Ischemia or Thrombosis," *New England Journal of Medicine*, 288:871–878, 1973.

"The First Enovid Report: Ad Hoc Committee for the Evaluation of a Possible Etiologic Relation with Thromboembolic Conditions," *Journal of New Drugs*, 3:201–211, 1963.

"Mortality Among Oral-Contraceptive Users: Royal College of General Practitioners' Oral Contraceptive Study," *Lancet*, 2:727–731, 1977.

"Oral Contraceptives and Cancer Risk," *Morbidity and Mortality Weekly Report*, 31,29:393–394, 1982.

Oral Contraceptives and Health, London, Royal College of General Practitioners. London: Pitman Publishing Corporation, 1974.

"Oral Contraception and Thrombo-Embolic Disease," *Journal of General Practitioners*, 13:267–279, 1967.

"Oral Contraceptives, Venous Thrombosis and Varicose Veins: Royal College of General Practitioners' Oral Contraceptive Study," *Journal of The Royal College of General Practitioners*, 28:393–398, 1978.

"Proposed Inserts Would List Birth Pill's Possible Benefits," *The New York Times*, November 25, 1985, p. B17.

Steroid Contraception and the Risk of Neoplasia, Technical Report Series no. 619, Geneva, World Health Organization, 1978.

"Study Linking Birth Control Pill to Breast Cancer Is Attacked," *The New York Times*, May 8, 1984, p. C3.

Close-Up (2)
TOXIC SHOCK SYNDROME

Bergdoll, M. S., Reiser, R. F., Cross, D. A., et al., "A New Staphylococcal Enterotoxin: Enterotoxin F, Associated With Toxic Shock Syndrome Staphylococcus Aureus Isolates," *Lancet*, 1:1017–1021, 1981.

Brody, J. E., "Doctors Urged to Watch For Toxic Shock in Wide Range of People," *The New York Times*, January 12, 1982, pp. C1, C4.

Brody, J. E., "Scientists Unraveling Mystery of Toxic Shock," *The New York Times*, April 26, 1983, pp. C1, C4.

Davis, J. P., Chesney, P. J., Ward, P. J., et al., "Toxic Shock Syndrome: Epidemiologic Features, Recurrence, Risk Factors and Prevention, *New England Journal of Medicine*, 303, 25:1429–1435, 1980.

Davis, J. P., Osterholm, M. T., Helms, C. M., et al., "Tri-State Toxic Shock Syndrome Study. II. Clinical and Laboratory Findings," *Journal of Infectious Diseases*, 145, 4:441–448, 1982.

Dolan, C., and Ingrassia, P., "Toxic Shock Victim Awarded $10.5 Million in Decision Against Johnson and Johnson," *The Wall Street Journal*, December 24, 1982, p. 22.

Glasgow, L. A., "Staphylococcal Infection in the Toxic-Shock Syndrome," (editorial) *New England Journal of Medicine*, 303, 25:1473–1474, 1980.

Harvey, M., Horwitz, R. I., and Feinstein, A. R., "Toxic Shock and Tampons: Evaluation of the Epidemiologic Evidence," *Journal of the American Medical Association*, 248, 7:840–846, 1982.

Hulka, B. S., "Tampons and Toxic Shock" (editorial), *Journal of the American Medical Association*, 248, 7:872–874, 1982.

Lawrence, S. V., "Disease Update: Toxic Shock Syndrome," *American College of Physicians Observer*, March 1984, p. 21.

Osterholm, M. T., Davis, J. P., Gibson, R. W., et al., "The State Toxic Shock Syndrome Study: I. Epidemiologic Findings," *Journal of Infectious Diseases*, 145:4, 431–440, 1982.

Robertson, N., "Toxic Shock," *The New York Times Magazine*, September 19, 1982, pp. 30–33, 109, 112, 116–117.

Schleck, W. F., Shands, K. N., Reingold, A. L., et al., "Risk Factors for Development of Toxic Shock Syndrome: Association With Tampon Use," *Journal of the American Medical Association*, 248, 7:835–839, 1982.

Schlievert, P. M., and Kelly, J. A., "Staphylococcal Pyrogenic Exotoxin Type C: Further Characterization," *Annals of Internal Medicine*, 96, 6 (part two):982–986, 1982.

Schlievert, P. M., Shands, B. P., Dan, G. P., et al., "Identification and Characterization of an Exotoxin from *Staphylococcus aureus* Associated with Toxic Shock Syndrome," *Journal of Infectious Diseases*, 143, 4:509–516, 1981.

Schmeck, H. M. "Scientists Isolate Gene Linked to Toxic Shock," *The New York Times*, October 20, 1982, p. A22.

Schrock, C. G., "Disease Alert," *Journal of the American Medical Association*, 243, 12:1231, 1980.

Senero, R., "Toxic Shock Cases Drop, But Disease Persists," *The New York Times*, May 1, 1982, p. 28.

Shands, K. N., Schmid, G. P., Dan, B. B., et al., "Toxic Shock Syndrome In Menstruating Women: Association with Tampon Use and Staphylococcal aureus and Clinical Features in 52 Cases," *New England Journal of Medicine*, 303, 25:1436–1442, 1980.

Todd, J., Fishaut, M., Kapral, F., et al., "Toxic Shock Syndrome Associated with Group I Staphylococci," *Lancet*, 2:1116–1118, 1978.

Wanda, K. A., "Tampons Can Be Harmful to Health," *The Nation*, January 2–9, 1982, pp. 1, 16–18.

"Data on Toxic Shock Syndrome, Tampon Revealed," *American Medical News*, December 23/30, 1983, p. 15.

"Follow-Up On Toxic Shock Syndrome—United States," *Morbidity and Mortality Weekly Report*, 29, 25:297–299, 1980.

"Follow-Up On Toxic Shock Syndrome," *Morbidity and Mortality Weekly Report*, 29, 37:441–445, 1980.

"Holdup of Toxic Shock Data Ends During Trial in Texas," *Journal of the American Medical Association*, 250, 24:3267–3269, 1983.

"Procter and Gamble Settles a Toxic Shock Suit," *The New York Times*, August 25, 1982, p. A16.

Toxic Shock Syndrome: Assessment of Current Information and Future Research Needs, Washington, D.C., Institute of Medicine Committee on Toxic Shock Syndrome, National Academy Press, 1982.

"The Toxic Shock Syndrome," A Conference Held on 20–22 November, 1981. Sponsored by the Institute of Medicine, National Academy of Sciences, *Annals of Internal Medicine* 96, 6 (part two):831–996, 1982.

"Toxic Shock Syndrome and the Vaginal Contraceptive Sponge," *Morbidity and Mortality Weekly Report*, 33, 4:43–44, 49, 1984.

"Toxic Shock Syndrome—United States," *Morbidity and Mortality Weekly Report,* 29, 20:229–230, 1980.

Close-Up (3)
X RAYS
Abrams, H. L., "The Overutilization of X Rays," *New England Journal of Medicine,* 300, 21:1213–1216, 1979.

Biddle, W., and Severo, R. D., "A Roundtable: Low-Level Dangers and Medicine's Responsibility—With Radiation, How Little Is Too Much?", *The New York Times,* September 26, 1982, p. E20.

Bailar, J. C., "Mammography—A Time for Caution," *Journal of the American Medical Association,* 237,10:997–998, 1977.

Bell, R. S., and Loop, J. W., "The Utility and Futility of Radiographic Skull Examination for Trauma," *New England Journal of Medicine,* 284, 5:236–239, 1971.

Black, M. M., Leis, H. P., and Kwon, S., "The Breast Cancer Controversy: A Natural Experiment," *Journal of the American Medical Association,* 237, 10:970–971, 1977.

Boffey, P. M., " 'Safe' Form of Radiation Arouses New Worry," *The New York Times,* August 2, 1983, p. C1.

Boffey, P. M., "Radiation Risk May Be Higher Than Thought," *The New York Times,* July 26, 1983, p. C1.

Boice, J. D., and Land, C. E., "Adult Leukemia Following Diagnostic X Rays? (Review of Report by Bross, Ball and Falen on a Tri-State Leukemia Survey)," *American Journal of Public Health,* 69, 2:137–145, 1979.

Broad, W., "Health Expert Finds Hazard of Radiation Worse Than Feared," *The New York Times,* March 18, 1983, p. A30.

Brody, J. E., "Cancer Group Urges X-Ray Breast Tests in Younger Women," *The New York Times,* August 3, 1983, pp. A1, B9.

Bross, I. D. J., Ball, M., and Falen, S., "A Dosage Response Curve for the One Rad Range: Adult Risks From Diagnostic Radiation," *American Journal of Public Health,* 69, 2:130–136, 1979.

Bross, I. D. J., and Natarajan, N., "Genetic Damage from Diagnostic Radiation," *Journal of the American Medical Association,* 237:2399–2401, 1977.

Bross, I. D. J., and Natarajan, N., "Leukemia From Low-Level Radiation: Identification of Susceptible Children," *New England Journal of Medicine,* 287:107–110, 1972.

Bull, J. W. D., and Zilkha, F. J., "Rationalizing Requests for X-Ray Films in Neurology," *British Medical Journal,* 4:569–570, 1968.

Feig, S. A., "Low-Dose Mammography: Application to Medical Practice," *Journal of the American Medical Association,* 242, 19:2107–2109, 1979.

Frost, S. B., Fearon, Z., and Hyman, H. H., *A Consumer's Guide to Evaluating Medical Technology*, New York, Consumer Commission on the Accreditation of Health Services, 1979.

Gofman, J., *Radiation and Human Health*. San Francisco: Sierra Club Books, 1982.

Gruson, L., "238 X-Ray Scanners Face Action by U.S. Over Radiation Risk," *The New York Times*, April 4, 1983, pp. A1, B8.

Gruson, L., "Technicare's CAT Scanner Woes," *The New York Times*, June 19, 1983, p. F4.

Gruson, L., "238 X-Ray Scanners Face Action by U.S. Over Radiation Risk," *The New York Times*, April 4, 1983, pp. A2, B8.

Hammerstein, G. R., Miller, D. W., and White, D. R., "Absorbed Radiation Dose in Mammography," *Radiology*, 140:483–490, 1983.

Hicks, M. J., Davis, J. R., Layton, J. M., and Present, A. J., "Sensitivity of Mammography and Physical Examination of the Breast for Detecting Breast Cancer," *Journal of the American Medical Association*, 242, 9:2080–2083, 1979.

Hutchison, G. B., Jablon, S., Land, C. E., and MacMahon, B., "Review of Report by Mancuso, Stewart and Kneale of Radiation Exposure of Hanford Workers," *Health Physics*, 37, 2:207–220, 1980.

Jablon, S., and Miller, R. W., "Army Technologists: 29-Year Follow-Up for Cause of Death," *Radiology*, 126:677–679, 1978.

Kassler, J., "Radiologists Urge Fewer Chest X Rays," *The New York Times*, November 23, 1983, p. C2.

Kopans, D. B., Meyer, J. E., and Sadowsky, N., "Breast Imaging," *New England Journal of Medicine*, 310, 15:960–967, 1984.

Lesnick, G. J., "Detection of Breast Cancer in Young Women," *Journal of the American Medical Association*, 237, 10:967–969, 1977.

MacMahon, B., "Prenatal X-Ray Exposure and Childhood Cancer," *Journal of the National Cancer Institute*, 28:1173–1191, 1962.

MacMahon, B., "Susceptibility to Radiation-Induced Leukemia," *New England Journal of Medicine*, 287:144–145, 1972.

Mancuso, T. F., Stewart, A., and Kneale, G., "Radiation Exposure of Hanford Workers Dying From Cancer and Other Causes," *Health Physics*, 33:369–385, 1977.

Marx, J. L., "Low-Level Radiation: Just How Bad Is It?", *Science*, 204:160–164, 1979.

Sanders, B. S., "Low-Level Radiation and Cancer Deaths," *Health Physics*, 34:521–538, 1978.

Stewart, A., Pennybacker, W., and Barber, R., "Adult Leukemias and Diagnostic X Rays," *British Medical Journal*, 2:882–890, 1962.

Stewart, A., Webb, J., and Hewitt, D., "A Survey of Childhood Malignancies," *British Medical Journal*, 1:1495–1508, 1958.

Swartz, H. M., "The Risks of Mammograms," *Journal of the American Medical Association*, 237, 10:965–966, 1977.

Advisory Committee on the Biological Effects of Ionizing Radiation: *The Effects on Populations of Exposure to Low Levels of Ionizing Radiation*, Washington, D.C., U.S. Government Printing Office, 1972.

Assessing the Efficacy and Safety of Medical Technologies, Office of Technology Assessment, U.S. Congress, Washington, D.C., U.S. Government Printing Office, September 1978, pp. 33–39.

"Breast Cancer: The Challenge of Early Detection," *The Harvard Medical School Health Letter*, VIII, 4:1–2, 5, 1983.

"Diagnostic and Screening Tests," *Health Facts*, IV, 2:1–8, 1980.

The Effects on Populations of Exposure on Low Levels of Ionizing Radiation, ("BEIR III" Report), National Academy of Sciences, Committee on the Biological Effects of Ionizing Radiation, Washington, D.C., U.S. Government Printing Office, 1980.

"Four Medical Advances: An Update," *Health Facts*, VII, 29:1–3, 1981.

Hearings, Subcommittee on Health and the Environment, Committee on Interstate and Foreign Commerce, House of Representatives, 95th Congress, *Effects of Radiation on Human Health*, Washington, D.C., U.S. Government Printing Office, 1979.

"Medical X Rays: What Your Doctor Doesn't Tell You," *Health Facts*, IV, 2:1–8, 1980.

"NIH/NCI Consensus Development Meeting on Breast Cancer Screening," *Preventive Medicine*, 7:269–278, 1978.

Chapter 5
GOVERNMENT GUARDIANS

Green, M., "Don't Just Do Something, Stand There," *The Village Voice*, March 15, 1983, pp. 9–12.

Hinds, M. de C., "Reagan's Drive to Cut Rules," *The New York Times*, January 22, 1982, p. A1.

Lavin, J. H., "A Fresh and Friendlier Wind Blows from the FDA, *Medical Economics*, August 9, 1982, pp. 47–48, 52, 56.

Tolchin, S., and Tolchin, M., *Dismantling America*. Boston: Houghton, Mifflin Co., 1984.

"Are Americans Getting the Government They Want?" *Common Cause*, May/June 1983.

Close-Up (1)
THE BATTLE OVER THE AIR BAG

Buss, D., "Small Cars May Save Fuel, But Cost Lives," *The Wall Street Journal*, April 27, 1982, p. 1.

Holusha, J., "What Deregulation Means for G.M.," *The New York Times*, November 1, 1981, p. F1.

Karr, A., "Death on the Road," *The Wall Street Journal*, May 4, 1982, p. 1.

Passell, P., "What's Holding Back Air Bags?", *The New York Times Magazine*, December 18, 1983, pp. 68–80.

Sloyan, P., "The Air Bag Lives—Abroad," *Newsday*, November 9, 1981, p. 113.

Will, G., "Driving Without Restraint," *The Washington Post*, April 19, 1977.

Passive Restraints for Automobile Occupants—A Closer Look, Report to the Congress by the General Accounting Office, Washington, D.C., July 27, 1979.

Passive Restraints: Ready When You Are, a booklet by Allstate Insurance Company, n.d.

Policy Options for Reducing the Motor Vehicle Crash Injury Cost Burden, Washington, D.C., Insurance Institute for Highway Safety, May 1981.

"Transcript of Nixon Meeting . . .", *Automotive Litigation Report*, November 18, 1982.

Close-Up (2)
FORMALDEHYDE POISONING

Kerr, P., "Debate on Safety of Urea Foam," *The New York Times*, February 25, 1982, p. C6.

Lewin, T., "Insulation Lawsuits Abound," *The New York Times*, May 25, 1982, p. D1.

Pittle, R., Sloan, D., Statler, S., Steorts, N., and Zagoria, S., Statements by (concerning UFFI ban), February 22, 1982.

Shabecoff, P., "E.P.A. Will Consider Regulation of Formaldehyde," *The New York Times*, May 19, 1984, p. 24.

"Ban of Urea-Formaldehyde Foam Insulation," by Consumer Product Safety Commission, 16 CFR Part 1306, *Federal Register*, April 2, 1982.

Formaldehyde: Questions and Answers, Scarsdale, New York, The Formaldehyde Institute, n.d.

"Petition to U.S. Consumer Product Safety Commission," Consumer Federation of America, Washington, D.C., August 1982.

Chapter 6
CALCULATING RISKS

Close-Up
THE CLEAN AIR DEBATE

Brown, M., *Laying Waste: The Poisoning of America by Toxic Chemicals*. New York: Pantheon Books, 1979.

Crouch, E., and Wilson, R., *Risk/Benefit Analysis*. Cambridge, Mass.: Ballinger Publishing Co., 1982.

Gordon, L., and Chapman, M., "Clean Air: A Right or a Luxury?", *Journal of Public Health Policy*, 3, 3:241–243, 1982.

Hacking, I., "Why Are You Scared?", *The New York Review of Books*, September 23, 1982, pp. 30–41.

Jones-Lee, M. W., *The Value of Life: An Economic Analysis*. Chicago: University of Chicago Press, 1976.

Layard, R., *Cost-Benefit Analysis*. New York: Penguin Books, 1974.

Mishan, E. J., *Cost-Benefit Analysis*. New York: Praeger, 1976.

Norman, G., "Risky Business," *Esquire*, November 1983, pp. 214–216.

Parish, R. M., "The Scope of Benefit-Cost Analysis," *Journal of the Economic Society of Australia and New Zealand*, 52:302–314, 1976.

Rowe, A., *An Anatomy of Risk*. New York: John Wiley and Sons, 1977.

Rowe, W. D., "Governmental Regulations of Societal Risks," *George Washington Law Review*, 45:944–968, 1977.

Shabecoff, P., "Reagan Order On Cost-Benefit," *The New York Times*, November 6, 1981, p. A18.

Shabecoff, P., "Ruckelshaus Said to Have Wanted Air Rules Eased," *The New York Times*, April 28, 1983, p. A15.

Starr, C., "Social Benefit Versus Technological Risk: What Is Our Society Willing to Pay for Safety?", *Science*, 165:1232–1238, 1969.

Stokey, F. J., and Zeckhauser, R., *A Primer for Policy Analysis*. New York: W. W. Norton, 1978.

Wines, M., "Automobile Bumper Standard Crumples as Cost-Benefit Analysis Falls Short," *National Journal*, January 23, 1982, pp. 145–149.

Wolcott, R., and Rose, A., *The Economic Effects of the Clear Air Act*, A Report for the Natural Resources Defense Council, Washington, D.C., March 1982.

"Breathing Easier," *Life*, May 1982, pp. 85–90.

"Tales from the Cost-Benefit Wonderland," *Consumer Reports*, June 1981, pp. 338–339.

Chapter 7
THE ULTIMATE RISK: THE GREENHOUSE EFFECT

Bernard, H., *The Greenhouse Effect.* New York: Harper/Colophon Books, 1981.

Hansen, J., Johnson, D., Lacis, A., et al., "Climate Impact of Increasing Carbon Dioxide," *Science,* 213:957–966, 1981.

Keerdoja, E., and Hager, M., "Is the Earth Getting Hotter?", *Newsweek,* October 31, 1983, p. 89.

Revelle, R., "Carbon Dioxide and World Climate," *Scientific American,* August 1982, pp. 35–42.

Shabecoff, P., "EPA Report Says Earth Will Heat Up Beginning in 1990's," *The New York Times,* October 18, 1983, p. 1.

Sullivan, W., "Experts Question Sea-Rise Theory: New Evidence Raising Doubts on Polar Ice Melting from Global Warming Trend," *The New York Times,* April 15, 1984, p. 15.

Sullivan, W., "Study Finds Warming Trend," *The New York Times,* August 22, 1981, p. A1.

Carbon Dioxide and Climate: A Scientific Assessment Report To The National Research Council, Washington, D.C., 1979.

Carbon Dioxide and Climate: A Second Assessment Report To The National Research Council, Washington, D.C., 1982.

Energy and Climate, Washington, D.C., National Academy of Sciences, 1977.

Understanding Climate Change, Washington, D.C., National Academy of Sciences, 1975.

Chapter 8
WHAT YOU CAN DO ABOUT HEALTH RISKS

Arnold, C. B., Kuller, L. H., and Greenlick, M. R., *Advances In Preventive Medicine,* volume 1. New York: Springer Publishing Company, 1981.

Blum, A., "Medical Activism" in *Health Promotion: Principles and Clinical Applications,* edited by R. B. Taylor. New York: Appleton-Century-Crofts, 1982, pp. 373–391.

Breslow, L., "Risk Factor Intervention for Health Maintenance," *Science,* 200, 26:908–912, 1978.

Breslow, L., and Somers, A. R., "The Lifetime Health Monitoring Program: A Practical Approach to Preventive Medicine," *New England Journal of Medicine,* 296, 11:601–608, 1977.

Brody, J. E., "Finding a Method to Reduce Stress," *The New York Times,* Februry 10, 1983, pp. C1, C10.

Frame, P. S., and Carlson, S. J., "A Critical Review of Periodic Health Screening Using Specific Screening Criteria," *Journal of Family Practice*, 2:29–36, 123–129, 189–194, 1975.

Faber, M. M., and Reinhardt, A. M., *Promoting Health Through Risk Reduction*. New York: Macmillan Publishing Co., Inc., 1982.

Farquhar, J. W., *The American Way of Life Need Not Be Hazardous to Your Health*. New York: W. W. Norton, 1978.

Farquhar, J. W., "Community Based Model of Lifestyle Intervention Trials," *American Journal of Epidemiology*, 108:103–111, 1978.

Fletcher, S. W., and Spitzer, W. O., "Approach of the Canadian Task Force to Periodic Health Examination," *Annals of Internal Medicine*, 92:253–254, 1980.

Knowles, J. H., *Doing Better and Feeling Worse: Health in the United States*. New York: W. W. Norton, 1977.

Murray, L., "Risk Factors: A Pound of Prevention," *American Health*, July/August 1982, p. 15.

Puska, P., "Recent Developments in the Field of Community Control of Cardiovascular Diseases in Finland," *WHO Meeting on Comprehensive Cardiovascular Control Programs*, Geneva, World Health Organization, 1977.

Shapiro, H. D., "Quit Smoking, and Cut Insurance Fees," *The New York Times*, September 4, 1983, p. F9.

Somers, A. R., *Promoting Health: Consumer Education and National Policy* (Report of the Task Force on Consumer Health Education to the National Conference on Preventive Medicine), Germantown, Maryland, Aspen Systems, 1976.

Spitzer, W. O. (chairman), "Report of the Task Force on the Periodic Health Examination," *Canadian Medical Association Journal*, 121:1193–1254, 1979.

Spitzer, W. O., and Brown, B. P., "Unanswered Questions About the Periodic Health Examination," *Annals of Internal Medicine*, 83:257–263, 1975.

"American Cancer Society Report on the Cancer-Related Health Checkup," *Cancer*, 30:194–240, 1980.

American College of Physicians, Medical Practice Committee, "Recommendations For Periodic Health Examinations," *American College of Physicians Observer*, December 1971, pp. 8, 11.

Conference On Health Promotion and Disease Prevention, volumes I and II, Washington, D.C., National Academy of Sciences Institute of Medicine, 1978.

"Final Meeting Held For Risk Factor Update Project," *Preventive Medicine Newsletter*, XXII,2:1, 4, 1982.

"Health Risk Appraisal—United States," *Morbidity and Mortality Weekly Report*, 30, 11:133–135, 1981.

Multiple Risk Factor Intervention Trial Research Group, "Multiple Risk Factor Intervention Trial," *Journal of the American Medical Association*, 248, 12:1465–1477, 1982.

Oslo Study Group, "MRFIT and the Oslo Study," *Journal of the American Medical Association*, 249, 7:893–894, 1983.

Preventive Medicine USA (Task Force Reports sponsored by the John E. Fogarty International Center for Advanced Study in the Health Sciences, National Institutes of Health and the American College of Preventive Medicine), New York, Prodist, 1976.

Index